T0279693

THE POWER OF PRIONS

THE POWER OF PRIONS

The Strange and Essential Proteins That Can Cause Alzheimer's, Parkinson's, and Other Diseases

MICHEL BRAHIC

PRINCETON UNIVERSITY PRESS
PRINCETON & OXFORD

Published by Princeton University Press
41 William Street, Princeton, New Jersey 08540
99 Banbury Road, Oxford OX2 6JX

press.princeton.edu

ISBN 9780691252384
ISBN (e-book) 9780691252407

British Library Cataloging-in-Publication Data is available

Editorial: Ingrid Gnerlich and Whitney Rauenhorst
Production Editorial: Mark Bellis
Text and Jacket Design: Katie Osborne
Production: Jacquie Poirier
Publicity: Matthew Taylor and Kate Farquhar-Thomson
Copyeditor: Maia Vaswani

Jacket Credit: VLA / SCIENCE PHOTO LIBRARY

This book has been composed in Arno Pro with Midnight Sans

Printed in the United States of America

10 9 8 7 6 5 4 3 2 1

To Lucie, Sula, Cora, Isla, and Marlo

CONTENTS

ILLUSTRATIONS

PREFACE

I was at my laboratory bench one morning in 1980 when a young neurologist walked in and declared that he had identified the agent of scrapie, a mysterious disease that caused behavioral changes, trembling, and eventually death in sheep and goats and left microscopic holes in their brains.

I had known Stanley Prusiner for some years and had been watching the progress of his laboratory in characterizing this highly unusual "microbe." I had also seen the tremendous amount of skepticism raised by his outlandish claim that it lacked genes, that it was an infectious protein. Stan's issue that morning was what to call this unique infectious agent. He had two candidates, he said: "piaf" and "prion." I forget what *piaf* stood for, but I said that the name was already taken by a very popular French singer. Fine, he said, in any case he preferred *prion*, a contraction of *protein* and *infection*. I agreed. What I did not say was that *prions* in my native French tongue means "let us pray," and that if he persisted with his claim of an infectious protein, he would need prayers. As we know, Stan was right, and for the discovery of prions he was awarded the Nobel Prize in 1997.[1]

1. Stanley Prusiner proposed the name "prion" for the scrapie agent in the article "Novel Proteinaceous Infectious Particles Cause Scrapie" (*Science* 216, no. 4542 [1982]:136–44).

Scrapie is an unusual brain disease of sheep, and its cause was a mystery until Stanley Prusiner discovered prions. But scrapie is not the only prion disease. Another, which was also mysterious and on top of that exotic, is kuru. Kuru was a human disease that was discovered and studied in the 1960s in the Highlands of New Guinea, where it was spreading among Aborigines practicing cannibalism. Kuru has a close cousin called Creutzfeldt-Jakob disease. The latter is present in all human populations around the world and, fortunately, is rare. Through the last few decades of the twentieth century, the list of prion diseases kept getting longer. Some may remember the "mad cow" epidemic of the 1980s and 1990s. And there is now the equivalent of mad cow disease in elk and some other wild animals. These are all brain diseases causing various motor and behavioral symptoms, including dementia. Probably the most curious symptom was observed in kuru, which was nicked named the "laughing death" because of the outbursts of laughing in patients in its terminal stages. Because these diseases all cause millions of tiny holes, called *vacuoles,* in the brain, they are grouped under the name *spongiform encephalopathies*—diseases that, under the microscope, give the brain the look of a sponge. All of them are caused by a protein called PrP, short for *prion protein.*

My interest in spongiform encephalopathies dates back to my medical training. I was attracted by neurology and neuroscience in general. When I was a student, the study of the biology of the brain, especially at the molecular level, was in its infancy. There was, however, a strong belief that neurological diseases were "experiments of nature," and that by deciphering their mechanisms one could learn something about the normal functioning of the brain. Already in the nineteenth century, pioneering neurologists such as Jean-Martin Charcot and Paul Broca in France and Karl Wernicke in Germany had mapped some essential functions, such as speech, to specific areas of the brain by carefully relating the symptoms of patients to the positions of lesions found in their brains at autopsy.

The fact that kuru and Creutzfeldt-Jakob disease caused dementia, the loss of higher brain functions, including memory, coupled with the mystery of their infectious nature, was highly intriguing and stimulating for a medical student in the 1970s. However, these diseases were difficult and expensive to study. Scrapie of sheep had been adapted to laboratory mice, but, just as in sheep, the animals did not shown symptoms and die for several months to more than a year after inoculation. As for kuru and Creutzfeldt-Jakob disease, they could be studied only in chimpanzees.

Faced with such difficulties, I turned instead to neurological diseases caused by viruses, including visna, a disease of sheep that resembled scrapie in some respects but was caused by a virus, which could be studied in the laboratory. I was working on visna when Stanley Prusiner entered the lab and wondered if he should call the scrapie agent *piaf* or *prion*. I remained a "classical" virologist and just a fascinated observer of the prion field until Ronald Melki, a colleague and a protein expert, told me of observations made by Patrick Brundin in Sweden and Jeffrey Kordower in the United States.

In 2008 both of these neurologists' teams had published articles suggesting that the spread of the lesions of Parkinson's disease in the brain could be explained if a protein called *alpha-synuclein* behaved like the prion protein of scrapie. This prompted Ronald Melki and me to test the hypothesis, and we experimentally confirmed the prediction of Brundin and Kordower. Such was my belated entry into the prion field. *Mieux vaut tard que jamais* (better late than never), as the French saying goes.

In fact, that a protein behaving as a prion, but different from PrP, was involved in Parkinson's disease vindicated Stanley Prusiner, who had intuited several years before that several common neurodegenerative diseases, including Alzheimer's and Parkinson's, could be caused by prion proteins. Let me emphasize right away that Alzheimer's, Parkinson's, and other neurodegenerative diseases are not infectious in the way scrapie is among sheep and goats. One cannot "catch" Alzheimer's disease by contact with patients.

But there are also "good" prions! An early and most surprising discovery was made at Columbia University in the laboratory of Eric Kandel. This group had been doing pioneering work on the molecular mechanism of memory using a simple model animal, a sea slug named *Aplysia*. After much work, they concluded that a protein with prion properties, but again different from PrP, was involved in memory storage in this animal. They went on to show that the same protein was involved in memory in the fruit fly and in mice. And there are more and more "good prion" proteins being discovered. It turns out that prion proteins appeared very early, maybe even at the very beginning of the evolution of life on this planet. We will discuss later how they help primitive organisms such as yeast to adapt to changes in their food environment. This field is young, fast-moving, and not devoid of controversy. The nomenclature is not yet settled. Some prefer to use the name *prion* only for PrP, the protein that causes kuru and the other spongiform encephalopathies. They call the others “prion-like proteins” or “prionoids.” Maybe it is time to resurrect *piaf* for those! But for the sake of simplicity, I will call all of them “prion proteins” or “prions.”

The main topic of this book is proteins and the nervous system. To make it accessible to readers with little knowledge of biology, the book will introduce the basics of cell biology, protein chemistry, and neuroscience. Biologists may want to skip some of these sections. Hopefully they will allow nonbiologists to understand the more specific topics, such as the role of synapse remodeling in long-term memory, and even to enjoy the beauty of protein folding and the model of self-templating.[2]

2. Several texts are available for readers unfamiliar with molecular biology. A useful one is *Molecular and Cell Biology for Dummies* by René Fester Kratz (2nd ed., Newark, NJ: John Wiley, 2020).

CHAPTER 1

Scrapie, Kuru, Cannibalism, and "Mad Cow" Disease

The story of prions goes back a long way. Scrapie, the disease of sheep, was already known in the eighteenth century by farmers in England and Germany. Sick animals would continuously rub themselves against fence posts, damaging their fleece, hence the name *scrapie*. They also showed other disturbing behaviors and some neurological symptoms such as tremors. Already in those days people knew that it had to be a brain disease. The animals were used mainly to produce wool, a valuable product, and the farmers were losing their income to scrapie. They dealt with their economic difficulties in the time-honored fashion by blaming somebody else for it, preferably someone from the south. In this case they accused Spaniards, from whom they had bought Merino sheep years before. With hindsight, they were probably right. Two centuries later, in the late 1930s, two French veterinarians, Jean Cuillé and Paul-Louis Chelle, published a series of articles that showed convincingly that scrapie was infectious, that it could be transmitted from sheep to sheep by intraocular injection, and that the incubation period was greater than one year. They also demonstrated that natural transmission occurred between animals housed together.

Sometime before this pioneering work, toward the end of the nineteenth century, Icelandic farmers had noticed that some of their sheep were coming down with a new disease, which they named *rida*.[1] In this case the farmers blamed a single ram that had been imported from Denmark some years before for bringing the disease to their flocks.[2] Needless to say, this origin was never proven. Rida was later shown to be scrapie. Then, in the 1930s, Iceland decided to import sheep from European countries and to cross them with their local animals to increase the quality of the wool and meat. This turned out to be a very unfortunate decision, as over the following years several previously unknown infectious diseases began to cause serious economic problems. One of them was visna, the neurological disease I was working on when Prusiner was looking for a name for the scrapie agent. The imported sheep had been quarantined and examined before they were released to farmers, but, because the diseases had unusually long incubation periods and protracted courses, infected animals went unnoticed. Furthermore, the imported sheep, having been exposed to their pathogens for centuries, were relatively resistant, whereas the Icelandic animals, which had been living in com-

1. *Rida* in Icelandic means "to tremble," "to stagger."

2. P. A. Palsson, "Rida in Iceland and Its Epidemiology," in *Slow Transmissible Diseases of the Nervous System*, edited by S. B. Prusiner and W. J. Hadlow (New York: Academic Press, 1979), 1:357.

plete isolation from the rest of the world, were highly susceptible.[3] These diseases were studied by Bjorn Sigurdsson, a talented veterinarian. He thought that they were all caused by viruses, including rida, and proposed calling them "slow viruses," to reflect the long incubation period and protracted course of the diseases.

Following Sigurdsson's work, slow viruses became an active field of research. When I became a virologist and started to work on visna, it was considered a frontier in virology. Visna was an interesting model for human multiple sclerosis. The virus looked like a tumor virus, but the disease was not a cancer. It became the prototype of a new group of viruses called *lentiviruses* (*lent* is French for "slow"), of which HIV became the most infamous member. But when Sigurdsson coined the name *slow viruses* he certainly did not realize that he would subject some of us to an easy joke that I heard far too many times at the beginning of my career: "There are no slow viruses, only slow virologists!" Colleagues are not always helpful.

Scrapie before Prions

After the pioneering work of Cuillé and Chelle in France, research on scrapie moved mainly to the United Kingdom and the United States. Some remarkable work was done despite enormous technical difficulties. Measuring the amount of the pathogen required using animals. Researchers injected

3. Infections play a major role in the evolution of species. They exert what is called a strong "selection pressure." Resistant individuals tend to reproduce more successfully than susceptible ones, and their presence in a population increases with time.

the animals with increasing dilutions of the sample and counted how many came down with disease and died from it. Several animals had to be injected for each dilution, and the amount of pathogen, its *titer*, was set as the dilution that killed 50 percent of the animals. This meant inoculating large numbers of sheep and waiting a year or more to get the result. Obviously, progress had to be slow. Important results were nevertheless obtained, including on the genetic susceptibility of various breeds of sheep and on the physicochemical properties of the agent. It was discovered that infectivity resisted treatments that inactivated all known parasites, bacteria, and viruses. This was a bewildering and important result.

A breakthrough occurred in 1961 when Richard Chandler at the Agricultural Research Council facility at Compton in the United Kingdom succeeded in adapting sheep scrapie to the laboratory mouse. This was a major advance. Infectivity assays still required observing the animals for a year or more, but they were small animals in cages, not sheep in barns. The work with mice took two directions. One was to study the mechanism of the disease, its *pathogenesis*: which organs besides the brain are infected and in what order, which cells carry the infection, and so on. These studies showed, for example, that when animals are inoculated in muscles, the organs of the immune system, in particular the spleen, become infected before the agent reaches the brain. The other direction was the characterization of the scrapie agent. Was it a virus or something else? The remarkable resistance of the agent to chemicals and radiation that inactivated viruses was puzzling. Its resistance to ultraviolet radiation was particularly unsettling. UV radiation interacts with nucleic acids, DNA and RNA. Resistance implied that nucleic acids were not essential for its infectivity. This was so unorthodox that it was met with enormous skepticism among biologists. Many of them tried to make models with viruses protected from the effect of UV radiation by unknown mechanisms. Even to this day, some—admittedly, a small minority—still cling to the idea that the scrapie agent contains a nucleic acid and that a virus is there that has just not yet been discovered.

Kuru in New Guinea

While the research on scrapie was ongoing in the United Kingdom and the United States, in 1954 Carlton Gajdusek, a pediatrician with a wide range of interests in biology, was spending the year in the laboratory of Sir Macfarlane Burnet at the Walter and Eliza Hall Institute of Medical Research in Australia. During a trip to New Guinea, he met Vincent Zigas, who was a medical doctor from Australia working in the Eastern Highlands of New Guinea. Zigas described to him a strange neurological disease that appeared to be epidemic in some isolated human populations in the Highlands, including a group that called themselves the Fore. The disease, which the Fore called *kuru,* was causing numerous deaths, mainly in women and children.[4]

4. The story of kuru and the work of Carleton Gajdusek in New Guinea are well described in P. P. Liberski, B. Sikorska, and P. Brown's "Kuru: The First Prion Disease" (chapter 12 in *Neurodegenerative Diseases,* edited by Shamin and Ahmad [New York: Springer, 2012]) and Liberski's "Kuru: A Journey Back in Time from Papua New Guinea to the Neanderthal's Extinction" (*Pathogens* 2 [2013]: 472–505). Vincent Zigas gave his own account of the early days of kuru research in *Laughing Death:*

Access to the Eastern Highlands of New Guinea is difficult. The mountains are high and separated by deep valleys with torrential rivers. As a result, the population is very fragmented, with small groups living in isolation. Gajdusek accompanied Zigas on one of his expeditions in the Highlands and became fascinated by the kuru disease and by the Fore people, who at the time were living in complete isolation, subsisting on hunting and gathering and farming root vegetables.[5] With great difficulty the two managed to set up an outpost where they could examine patients and work out the history of the disease, its *epidemiology*. They published their first findings in 1957. Gajdusek and Zigas worked under extreme physical hardship and amid great personal danger. The early studies of kuru would not have been possible without the unusual and extremely strong personality of Carleton Gajdusek. Sir Macfarlane Burnet, in whose laboratory Gajdusek was working at the time, gave this description of him: "I had heard that the only way to handle him was to kick him in the tail, hard. Somebody else told me he was fine but there just wasn't anything human

The Untold Story of Kuru (Clifton, NJ: Humana, 1990). Michael Alpers has written an important and enjoyable paper that gives all sorts of information on the Fore people and other kuru-affected groups, and on the early work done in the 1960s by him, Gajdusek, and others: "The Epidemiology of Kuru: Monitoring the Epidemic from Its Peak to Its End," *Philosophical Transactions of the Royal Society B* 363 (2008):3707–13.

5. Carleton Gajdusek was a prolific writer. No matter the circumstances, he would always find time to keep a daily account of his activities in his journal. A selection of notes taken during the early days of his work on kuru has been published: D. Carleton Gajdusek, *Kuru: Early Letters and Field-Notes from the Collection of D. Carleton Gajdusek*, edited by J. Farquhar and Gajdusek (New York: Raven 1981).

about him. My own summing up was that he had an intelligence quotient up in the 180s and the emotional immaturity of a 15-year-old. He is completely self-centered, thick-skinned, and inconsiderate, but equally won't let danger, physical difficulty or other people's feelings interfere with what he wants to do."[6]

In their outpost, Gajdusek and Zigas managed to perform autopsies and to ship brain samples to neuropathologists at the National Institutes of Health (NIH) in Bethesda and elsewhere. Later, when Gajdusek moved to the NIH, his laboratory worked almost exclusively on kuru. However, for many years he still spent extended periods of time living and trekking in the Highlands of New Guinea.

In 1959, Klatzo, Gajdusek, and Zigas published a lengthy description of the brain lesions in kuru.[7] They described widespread loss of neurons and the proliferation of glial cells called *astrocytes*. They compared these lesions with those of other neurological diseases and concluded that the only resemblance was with Creutzfeldt-Jakob disease, a rare human disease of unknown cause. They discarded the possibility of a viral infection because of the absence of infiltration by cells of the immune system, and the negative results of their inoculations to laboratory animals.

6. Quoted in Jay Ingram, *Fatal Flaws: How a Misfolded Protein Baffled Scientists and Changed the Way We Look at the Brain* (New Haven, CT: Yale University Press, 2013), 18.

7. I. Klatzo, D. C. Gajdusek, and V. Zigas, "Pathology of Kuru," *Laboratory Investigation* 8 (1959):799–847.

Kuru and Scrapie

The story of prions took another serendipitous turn a few years later. William Hadlow, a veterinarian from the NIH Rocky Mountain Laboratories, happened to be in England. He was alerted by a colleague to the presence of an exhibit in London about New Guinea, kuru, and Gajdusek's findings. Hadlow was interested in neuropathology and was an expert on scrapie. He went to the exhibit and was struck by the pathology micrographs taken by Gajdusek and his coworkers, which showed microscopic holes inside neurons and in the tissue between them, not mentioned by Klatzo, Gajdusek, and Zigas in their 1959 article. Hadlow thought that they closely resembled those observed in scrapie.

Following his visit to London, Hadlow wrote a short letter to the *Lancet* pointing out the similarities between the lesions of scrapie and those of kuru. He suggested inoculating nonhuman primates to determine if the disease was transmissible, possibly with a long incubation period like that of scrapie in sheep. The letter was a turning point. It prompted Gajdusek and his associate Joe Gibbs to perform a new round of inoculations, including of chimpanzees. As predicted by Hadlow, the chimpanzees came down with a kuru-like disease, but only one to two years after inoculation, depending on the animal. This was soon followed by the transmission of Creutzfeldt-Jakob disease to chimpanzees, again by Gibbs in the Gajdusek's laboratory. For these discoveries Carlton Gajdusek was awarded the Nobel Prize in Medicine in 1976.

The numerous microscopic holes in the brains of sheep with scrapie and humans with kuru, Creutzfeldt-Jakob disease,

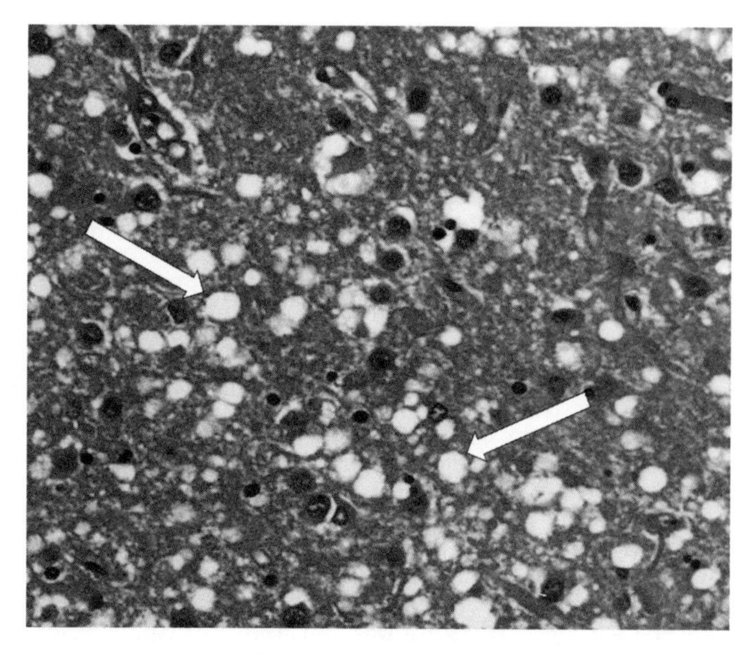

FIGURE 1.1. Section of a brain sample from a patient who died of Creutzfeldt-Jakob disease, as seen under the microscope. The arrows indicate vacuoles. The large number of vacuoles gives the brain the look of a sponge. From *Neuropathology Simplified: A Guide for Clinicians and Neuroscientists* by David A. Hilton and Aditya G. Shivane; © 2015 Springer International Publishing Switzerland

and some other diseases are called *vacuoles* (figure 1.1). These vacuoles, which are mainly located inside neurons, make parts of the brain look like a sponge under the microscope, hence the name *transmissible spongiform encephalopathies* (TSEs) now given to this group of diseases.

Veterinarians had shown that in the field scrapie was transmitted from animal to animal by the oral route, through eating contaminated grass and the placenta of ewes that had just delivered. Kuru could be transmitted in the laboratory to chimpanzees by intracerebral inoculation. But how was it transmitted among the Fore in New Guinea? Because of their extreme isolation, the populations of the Eastern Highlands of New Guinea had attracted the attention of anthropologists, among them Robert Glasse and Shirley Lindenbaum.[8] During their field studies with the Fore tribes, they observed that the geographic spread of kuru coincided with the practice of cannibalism. Deceased Fore were dissected, and their organs, including the brain, were cooked and eaten. This is commonly referred to as *ritual cannibalism* in the scientific literature, with the idea that the practice had something to do with the rebirth of the soul of the deceased in future generations. Shirley Lindenbaum, who studied cannibalism in the Fore group and was the first to suggest that kuru was transmitted by cannibalism, contests the term "ritual" and asserts that cannibalism was adopted to provide women and growing children with protein after the Fore became farmers growing root vegetables. Animal meat, from hunting, was reserved for men.

Even though scrapie was known to be transmitted by the oral route, the transmission of kuru by cannibalism was met at first with skepticism by scientists. However, it has been confirmed by the fact that no case of kuru has been observed in people born after the cessation of cannibalism. It is not possible to precisely determine the origin of the kuru epidemic. One can speculate that it began with a sporadic case of Creutzfeldt-Jakob

8. Shirley Lindenbaum wrote an account of her work with kuru: *Kuru Sorcery: Disease and Danger in the New Guinea Highlands* (London: Routledge, 2016).

disease in the Fore population. The consumption of brain tissue from this initial case and the common practice at the time of cannibalism by women and children may have triggered the epidemic.

Creutzfeldt-Jakob Disease

German neurologists Hans G. Creutzfeldt and Alfons M. Jakob first described the disease that bears their name in the 1920s. The disease is rare; its incidence is one to two cases per million per year. The symptoms consist of dementia with memory loss, hallucinations, and involuntary muscle contractions known as *myoclonus*. It is invariably fatal within a few months to a year, and there is no cure or preventive measure. The pathology of the disease is typical of the other spongiform encephalopathies, including scrapie and kuru. We already mentioned that the disease was transmitted to chimpanzees by intracerebral inoculation in the laboratory of Carleton Gajdusek.

There are rare familial cases of the already-rare Creutzfeldt-Jakob disease. Familial cases of diseases can be a great help to scientists. If geneticists find a mutation that is present in all cases in the family, and if the mutation is transmitted from generation to generation according to Mendel's laws of heredity, one can be almost certain that the product of the mutated gene is causing the disease. Despite its rarity, several families with Creutzfeldt-Jakob disease have been identified across the world. In all of them geneticists have found that the gene responsible for disease was *PRNP*, which is the gene that codes for the PrP protein. We will see later in the book how Stanley Prusiner showed that PrP is the agent of scrapie.

Unfortunately, Creutzfeldt-Jakob disease has occasionally been transmitted accidentally from human to human. The first case was reported in 1974, in a woman who had received a corneal transplant. She developed Creutzfeldt-Jakob disease two years later. It was later determined that the donor of the cornea had died from a neurological disease, which at autopsy turned out to be Creutzfeldt-Jakob disease.

Some years later, two cases were caused by neurosurgery for severe epilepsy. To limit the damage caused by the surgery, surgeons located the area to be removed by recording its abnormal electric activity with electrodes. The electrodes in these cases were sterilized in between patients with alcohol and formaldehyde at concentrations that kill all known bacteria, viruses, and parasites. A patient who had undergone neurosurgery was diagnosed with Creutzfeldt-Jakob disease sometime later. Two other patients for whom the same electrodes were used, after sterilization, also came down with Creutzfeldt-Jakob disease, fifteen and eighteen months respectively after surgery. We now know that the Creutzfeldt-Jakob prion is extremely resistant to sterilizing chemicals, including ethanol and formaldehyde at the concentrations used at the time. Transmission of Creutzfeldt-Jakob disease has also occurred with some batches of dura mater used during neurosurgery. Such accidental transmission is now prevented by measures that take the resistance of the infectious agent into account.

The most dramatic contamination happened with the use of human growth hormone. Growth hormone is a protein made by the pituitary gland, a small gland at the base of the brain. A deficit in growth hormone causes dwarfism, short stature, which can be prevented by treating children with the hormone. For many years, human growth hormone was extracted from pituitary glands obtained at autopsy. Then, in 1985 the Department of Health and Human Services in the United States was alerted to three cases of Creutzfeldt-Jakob disease in young men treated with human growth hormone for dwarfism. This prompted an investigation, also undertaken in several other countries, into the incidence of Creutzfeldt-Jakob disease among growth-hormone-treated patients. The conclusions were dire. In the United States, 35 cases were found, 80 in the United Kingdom, and 123 in France. The obvious conclusion was that some of the autopsies to obtain pituitary glands had been performed on people with undiagnosed Creutzfeldt-Jakob disease; individuals who died of other causes but had early lesions of Creutzfeldt-Jakob disease in their brain. The amount of growth hormone in the pituitary of elderly individuals is very small. Therefore, the hormone was purified from batches of many pituitaries. A single case of Creutzfeldt-Jakob disease among the donors was sufficient to contaminate the whole batch of hormone, and therefore several patients. Fortunately, this tragedy was followed by the development of synthetic human growth hormone made by genetic engineering. No case of Creutzfeldt-Jakob disease has been reported since the use of synthetic hormone.

The "Mad Cow" Epidemic: Variant Creutzfeldt-Jakob Disease

In the late 1980s, a new disease of cattle appeared in Britain, nicknamed "mad cow" disease by the press. Veterinarians determined that the cows were dying from a neurological disease and that the lesions in the brain were like those of scrapie in sheep. They named the disease *bovine spongiform encephalopathy* (BSE). In the years following, the disease appeared in several other countries around the world. The UK government knew that scrapie was transmitted by sheep eating contaminated food. They traced the appearance of BSE to a change in the methods of preparation of cattle food supplements from recycled livestock carcasses, a cost-cutting measure. They destroyed the stocks of these food supplements and began the systematic culling of all animals in herds with one or more cases of BSE, with burning of carcasses to destroy the infective agent. The cost to the economy was enormous. Over four million cattle were destroyed to eliminate the disease from the United Kingdom.

For me, "mad cow" disease was the occasion of a humiliating episode. I was not working on spongiform encephalopathies, but I had a keen interest in the field, and my laboratory at the Pasteur Institute in Paris was called the Slow Virus Unit. We were a nice group of scientists, including many students, and we did not mind being called "slow virologists" by our colleagues. I received several calls from journalists at the beginning of the BSE epidemic, many of them asking about the risk of contracting Creutzfeldt-Jakob disease through eating beef. People were worried.

I told all of them that there was absolutely no reason to stop eating beef. Scrapie had been present in Britain and the rest of Europe for a long time. It had a long incubation period. There was no way to screen for asymptomatic animals before they were sent to the slaughterhouse. Therefore, scrapie must have entered the human food chain long ago, and there was no evidence of a link between eating lamb and Creutzfeldt-Jakob disease. Scientists had done a great deal of epidemiological work, including surveying Libyan shepherds who eat sheep's eyes as a delicacy. No evidence was found of transmission to humans. Period. Besides, when I was in school, we had sheep's brains for lunch quite regularly, and I was fine. So that was that.

Unfortunately, around 1996–97, ten years after the peak of BSE, UK neurologists noticed cases of Creutzfeldt-Jakob disease with unusual brain lesions. They called these cases *variant Creutzfeldt-Jakob*, or vCJD. The number of cases of vCJD peaked around the year 2000 and then diminished. Over the same time period, the incidence of classical Creutzfeldt-Jakob disease remained the same as ever, between one and two new cases per million individuals every year. Variant Creutzfeldt-Jakob was also observed in other countries, but the majority of cases, more than two hundred, were in the United Kingdom, where "mad cow" disease had been much more prevalent than in the rest of the world.

Variant Creutzfeldt-Jakob disease illustrates the role of genetic susceptibility in prion diseases. The prion protein responsible for spongiform encephalopathies, PrP, is 230 amino acids long (we will discuss in the next chapter how proteins are made up of small molecules called amino acids). Amino acid 129 is either a methionine or a valine, depending on the genetic background of the individual. We all have two copies of each of our genes, one copy on each chromosome of a pair of chromosomes. For some people, both copies of the gene that codes for the PrP protein have a methionine at position 129 (40% of

Caucasians),[9] for others both have a valine (10% of Caucasians), and for still others one copy of the gene has methionine and the other has valine (50% of Caucasians). It turned out that virtually all the patients with vCJD had methionine on both chromosomes. This is a typical example of genetic susceptibility to a transmissible disease. The mechanism is not understood in the case of vCJD, but a hypothesis will be discussed in chapter 2.

What did I overlook when I was asked by journalists about the risk of BSE infecting humans? Pathogens have what is called a *host range*. For some, it is wide—they infect many different species. Others have a narrow host range. For example, measles virus infects only humans—as far as we know, of course. The PrP prion as known at the time had a narrow host range. Scrapie prion infects only sheep and goats. Careful epidemiological studies have not found evidence of transmission of scrapie to humans through eating lamb, nor even sheep's brains. The kuru and Creutzfeldt-Jakob prions are restricted to humans and some nonhuman primates, including chimpanzees. However, what I overlooked was that goat and sheep scrapie had been "adapted" to mice and hamsters on a few occasions in the 1960s and 1970s, often after several blind passages from mouse to mouse or hamster to hamster. Therefore, the possibility of a change in host range exists, even with the PrP prion.

We do not know by what mechanism a prion can be "adapted" to a new host. But we can build hypotheses based on what we learned about the mechanism of prion multiplication. This also will be discussed in chapter 2.

9. Virtually all the vCJD patients were in Britain and France and were Caucasian.

An Infectious Agent Made of Just One Protein

There are different personal styles among scientists. Some may trust their intuition and proceed to test a bold hypothesis unsupported by preliminary evidence. Others prefer to take a Cartesian approach, starting with what is already known for sure and going one step at a time. Of course, in most cases, progress is made through a mixture of both attitudes, and serendipity can play a big part in getting to the result. When Stanley Prusiner decided to identify the agent of scrapie, he opted for a step-by-step, rational approach. He knew, from the work of others, where to look for it—in the brain—and that it most likely contained proteins but possibly no DNA or RNA. That was not a lot of information.

To purify a component from animal tissue, scientists perform what they call *fractionations*. They start with a complex mixture, such as a piece of brain that has been homogenized in a blender, and try to separate it into its component parts, or *fractions*, keeping track of where the product they want to purify ends up in the process. They may place their mixture in a tube and spin it in a centrifuge, which can spin the tube at various precise speeds. At the right speed, one hopes to separate the product that one wants to purify between either the pellet at the bottom of the tube or the *supernatant*, the liquid at the top. For each speed tested, one needs to find out where the product is, in the pellet or the supernatant, and how much of it is in both. There is nothing more depressing than finding that there is just as much in both. This means that the purification step has achieved nothing. And, in fact, it was often the case for people trying to purify scrapie. Remember that the assay to detect the agent and quantify it took a year or more, because the only way to do it was to inoculate animals with various dilutions of the material and wait until they got sick and died. No wonder results were coming slowly.

Faced with these difficulties, Stanley Prusiner realized that he needed a starting material that was as rich in the agent as possible, and an assay that was as rapid as possible. After a series of attempts, which are described in detail in his autobiography *Madness and Memory*,[10] he settled on using hamsters instead of mice. The amount of scrapie agent in hamster brain was especially high, and, crucially, with hamsters he could speed up the assay because their disease was more rapid than that in laboratory mice. Furthermore, he devised an assay that did not require waiting until all infected animals had died. In preliminary experiments he determined that measuring the length of time between inoculation and the appearance of the first signs of disease gave an accurate titer of the agent. Using hamsters and a relatively fast assay paved the way to success. The only drawback, but a serious one, was that buying, housing, and observing daily a large number of hamsters was a lot more expensive than buying and housing mice.

At the end of a series of purification experiments, Prusiner concluded that the purest specimen he could obtain contained only protein, with one prominent one. Agents that damage proteins diminished or eliminated the infectivity. Agents acting on DNA and RNA had no effect on infectivity. It looked as though the scrapie agent was made of protein, possibly

10. Stanley B. Prusiner, *Madness and Memory* (New Haven, CT: Yale University Press, 2014).

only one type, and that nucleic acids were not required for infectivity. This is the point where he started looking for a name for the agent and came up with *prion*—a portmanteau from protein and infection.

The identification of the prion protein, abbreviated to PrP, took more time. Prusiner needed the help of molecular biologists. They purified the PrP protein further, were able to determine the sequence of its amino acids, and from there, with the help of more molecular biologists, they determined that the protein was encoded by one of the animal's genes. It was not a foreign protein, not a protein brought in by a microbe. They sequenced the gene coding for the protein and finally were able to obtain mice whose PrP gene had been eliminated by genetic engineering. Remarkably, these mice were totally resistant to inoculation with mouse scrapie. This was of course an important result, showing that the mouse PrP protein was required for the infection. However, it did not prove that the PrP protein by itself caused the disease; it only showed that the gene coding for the protein was needed. This could have been the case if, for example, the PrP protein had been the receptor for a scrapie virus. If there is no receptor for them to bind to, viruses cannot infect cells and cannot cause disease. Eventually, and more recently, Jiyan Ma at East China Normal University in Shanghai and Witold Surewicz from Case Western University in Cleveland, Ohio, showed that PrP prions obtained by genetic engineering could cause scrapie in mice and hamsters. The heretical protein-only hypothesis has been vindicated.[11]

11. This is only a summary of much research that led to the prion concept. Besides those mentioned, other researchers made essential contributions to the

But not for everybody. There are still a few biologists, including Laura Manuelidis at Yale University, who claim that spongiform encephalopathies are caused by a virus that has not yet been discovered, and that the PrP prion protein is a factor in the disease, or could be a consequence of the infection but not its cause. Skepticism is always welcome in science. Dogmas are dangerous, and one should also be wary of fashion. However, at present the overwhelming evidence is in favor of the protein-only original hypothesis.

But you may wonder, since PrP is a protein present in everybody's brain, why do only a few individuals come down with a dreadful spongiform encephalopathy? How can a normal protein in the brain suddenly, without any mutation, turn into a deadly pathogen? This is indeed a very good question. The next chapter will explain this. It requires first giving some background information on proteins and how they fold to acquire a three-dimensional shape.

To Recap

Scrapie is a disease of sheep known of since the eighteenth century. It can be transmitted to healthy animals by inoculation with brain extracts from sick ones. Studying scrapie was difficult because the incubation period can be more than a year, and because sheep are not laboratory animals. Scrapie

---

discovery of prions. Adriano Aguzzi, a prion expert, mentions several of them in his article "Prion Science and Its Unsung Heroes" (*Science* 383 [2024], https://doi.org/10.1126/science.adn94).

was transmitted to laboratory mice in the 1960s, making experimental work easier.

The scrapie agent is remarkably resistant to the chemical and physical agents that inactivate all known microbes. Resistance to UV radiation implies that the agent does not contain DNA or RNA.

Carleton Gajdusek and Vincent Zigas studied kuru in the Highlands of New Guinea. William Hadlow pointed out that the brain lesions of kuru and scrapie looked alike. Kuru was transmitted to chimpanzees by intracerebral inoculation. The incubation period was longer than one year. Kuru was transmitted in New Guinea by the practice of cannibalism.

The brains of scrapie-infected sheep and kuru patients both showed numerous microscopic holes, called *vacuoles*, which give the tissue the appearance of a sponge, hence the name *transmissible spongiform encephalopathies* (TSEs) given to this group of diseases.

Creutzfeldt-Jakob disease is a rare human TSE, which was accidentally transmitted to recipients during neurosurgery and through the administration of growth hormone extracted from human cadavers. The "mad cow" epidemic was a variant of scrapie that spread among cattle fed recycled livestock carcasses.

Stanley Prusiner purified the scrapie agent and showed that it was made of a single protein called PrP. PrP is a host protein; it does not come from a microbe. PrP became the prototype prion protein. It causes all spongiform encephalopathies: scrapie, "mad cow" disease, and several rare human diseases including kuru and Creutzfeldt-Jacob disease.

CHAPTER 2

The Life of Proteins

PRION PROTEINS ARE SPECIAL

After water, proteins are the second-most-abundant component of the human body.[1] They account for approximately 20 percent of our body weight. They are everywhere, both inside and outside our cells. They come in all sorts of shapes and perform a host of different functions. Although we do not have an exact number, it is estimated that there are somewhere between one and two hundred thousand different proteins in our body.

Proteins are very diverse in size and shape. Some are like little globules, others like rods, and many take a form somewhere on the spectrum in between. Proteins can be big or small, but even large ones are only a few nanometers in size (a nanometer is a millionth of a millimeter). Therefore, a single

1. There are several texts that might be useful to readers unfamiliar with molecular biology. One of them is *Molecular and Cell Biology for Dummies* by René Fester Kratz (2nd ed., Newark, NJ: John Wiley, 2020).

protein cannot be seen under a light microscope, nor even a standard electron microscope.

Some proteins are soluble and float in body fluids; for example, the antibodies that circulate in our blood are proteins. Inside cells, a milieu with a consistency akin to egg white, most proteins are not free-floating. Rather, they form complex assemblies, which can reach large sizes. For example, the scaffolding that gives cells and tissues their specific shapes is made of protein. Protein assemblies inside cells are not static—they form, disperse, and form again according to need. Some assemblies form very complex micromachines, such as *ribosomes* (of which more later), which make proteins. Ribosomes make proteins, but they are also *made of* protein—eighty different proteins—together with three different kinds of RNA molecules. They are large enough to be seen as dots with an electron microscope. Many proteins are enzymes, which catalyze chemical reactions that could not happen at body temperature without them. There would be no life without enzymes.

The Structure of Proteins

Proteins can be compared to chains of beads of different colors. In proteins the beads are small molecules called *amino acids*. There are twenty different kinds of amino acids—twenty kinds of beads, each with a different color. The order of the amino acids in the chain is called the *sequence* of the protein. Each of the hundred to two hundred thousand different proteins in our body has a different sequence of amino acids. Their lengths and sizes are also different. It is the differences in sequence and in length among proteins that create their enormous diversity.

In a protein, the chain of amino acids is not loose and flexible. Unlike a necklace made of beads, the chain of amino acids folds in a very precise way (we will discuss folding in more detail below), which gives each protein its own particular shape. To study and understand the way a protein functions, it is essential to know its shape—its three-dimensional structure—down to the position of every atom. There are techniques to determine this structure. One consists in making crystals of the pure protein and shining X-rays on the crystal. From the diffraction pattern obtained, it is possible to determine the structure of the protein. X-ray diffraction gives very precise results; the position of each atom can be accurately determined. But it is slow and cumbersome. It may take months or even years to find the conditions under which a given protein will form crystals. Some won't crystalize at all. In some cases, crystallization alters the structure of the protein. Sometimes the tiny crystals are very fragile and may shatter under the X-ray beam. Still, X-ray diffraction is the gold standard for elucidating protein structure.

Another technique, which was developed more recently, is called cryo-electron microscopy. Its big advantage is that it does not require making crystals. The solution containing the pure protein is frozen so quickly that the water surrounding the protein molecules does not have time to form ice crystals. The protein molecules, which are randomly oriented in the frozen sample, are imaged in a special and very powerful electron microscope. Because the protein molecules in the sample are oriented at random, the images show the protein at all possible angles. The trick consists in using the power of computers to sort out and align the thousands of two-dimensional images that correspond to the protein seen at a given angle. Because there are so many images of the same object at the same angle, the computer can produce a sharp, refined image. This process is repeated for all possible angles, allowing the computer to construct a high-resolution image of the molecule in three dimensions.

Some general rules governing the structure and the folding of proteins emerged from studies done using X-ray crystallography. The most important one is the existence of common folding patterns called *alpha helices* and *beta strands* (figure 2.1). Depending on the nature and order of the amino acids, some parts of a protein may be able to twist and form a helix. These are called alpha helices; they come in various lengths and are common in proteins. Figure 2.1 shows a protein with seven alpha helices.

Beta strands are very different from alpha helices. They consist in stretches of four to ten amino acids that form a sort of rigid zigzag. What determines whether a portion of a protein forms a beta strand is, again, the nature and the order (sequence) of amino acids in that part of the protein. By itself, a beta strand is not very stable. It is made stable if the zigzag interacts with other parts of the protein. A common sort of interaction is called a *beta sheet*. This forms when two beta strands, which can be far apart along the chain, stick to each other a bit like the two sides of a zipper. In some cases, several beta strands may align and stick, forming a pile of beta strands, a kind of multilayered sandwich (see figure 2.4C).

Together, alpha helices and beta strands form a sort of scaffold for the protein. Because a protein is a continuous string of amino acids, alpha helices and beta strands are connected

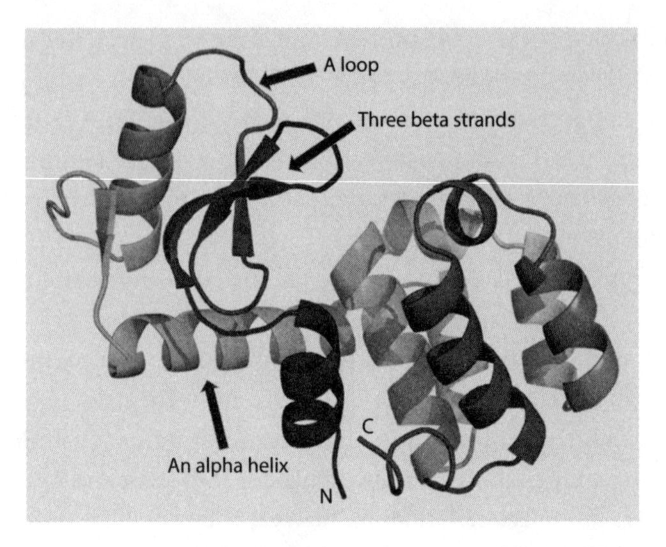

FIGURE 2.1. Computer-generated representation of the scaffolding of a protein, the T4 lysozyme. Seven alpha helices, four beta strands, and several loops are clearly visible. N and C indicate the two ends of the chain of amino acids. The 164 amino acids that make this protein are not represented. The drawing shows only the overall shape taken by the chain of amino acids after folding. Modified from *Exploring Protein Structure: Principles and Practice* by Tim Skern; © 2018 Springer International Publishing AG

to one another by other segments of the proteins, called *loops*. The structure, the overall shape, of the loops and their chemical properties are highly variable. This variability allows proteins to perform their many different functions, such as enzymes performing specific tasks, or antibodies that can recognize and bind to particular viruses. These functions usually require binding very specifically to other molecules, including to other proteins. This binding is sometimes compared to the interaction of a key with its lock, because of the precise shapes of key and lock. Besides its shape, the binding of a protein to a partner also involves other local properties, such as electric charge and hydrophobicity. Their loops of different lengths and different amino-acid compositions give proteins an infinite number of possibilities to form such very specific sites on their surface.

How Are Proteins Made by Cells?

Cells are the basic unit of all life-forms on the planet. All our tissues and organs are made of cells. Neurons, for example, are cells in our brain. Cells come in many shapes and perform different functions, but all share a basic common structure, which we will briefly review. We will deal here only with eukaryotic cells—cells that have a nucleus. Prokaryotic cells, such as bacteria, are a bit different.

A cell is a structure limited by a continuous membrane that separates it from its environment (figure 2.2). Membranes are made of many molecules of lipid, molecules of fat with two parts, one part that repels water and one part that doesn't. In the membrane, the parts of the lipid molecules that do not mix with water stick to one another and form a water-repellant layer. Water, and molecules in solution in water, cannot diffuse in and out of the cell because of this layer. The parts of the lipid molecules that can mix with water are situated on both sides of this impermeable layer. The cell is like a bag, made of an impermeable membrane, filled with all sorts of proteins and other molecules.

The membrane that limits a cell is called the *plasma membrane*. Inside cells one finds many other smaller compartments

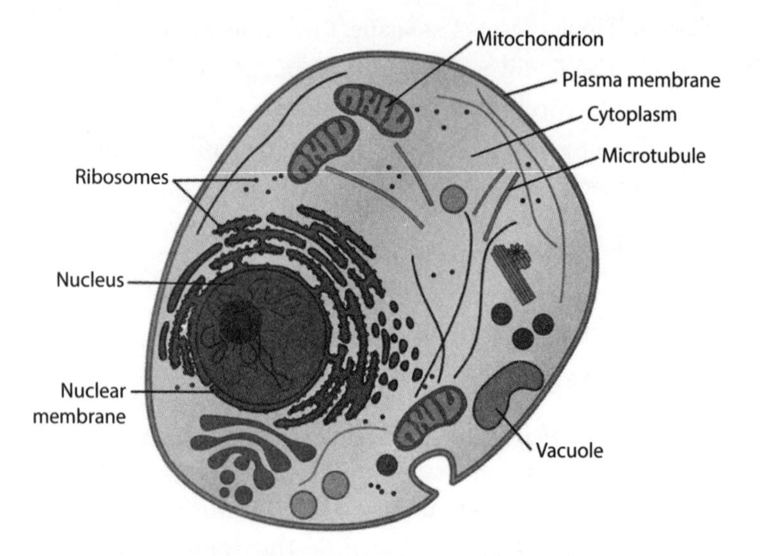

FIGURE 2.2. Diagram of a typical animal cell. The cell contains many different organelles, but only those mentioned in the book are labeled. Vladimir Ischuk/Shutterstock

also limited by membranes. For example, the nucleus, which contains the chromosomes and their DNA, is separated from the cytoplasm by the *nuclear membrane*. You may wonder how all this can work if each cell and each compartment within a cell is isolated from the rest by membranes. It works because there are many channels across membranes, which allow exchange of molecules between the compartments and between the cell and its environment.

These channels are made of proteins, with a transmembrane part embedded in the water-repellant lipid layer and two other parts, one on the inside and one on the outside. These channels are not just holes in the membranes; they are very specific in what they allow in or out. Some of these channels, especially in neurons, can be open or closed depending on the conditions. Besides channel proteins, membranes have other proteins attached to them. Some cross the membrane lipid layer and have one part inside and another part outside. Others do not cross the lipid layer and are located entirely on the outside or the inside of the membrane.

All proteins are synthesized inside cells by a mechanism that has been studied in detail and is summarized in figure 2.3. Proteins are the products of genes. Genes are segments of DNA in chromosomes. Like a protein, a gene is a chain of beads, but this time there are only four different kinds of beads, called *nucleotides*, which are represented by the letters A, T, G, and C. The gene contains the blueprint of the protein—its sequence of amino acids—written in code using the four different nucleotides. In the gene, each amino acid is represented by a group of three nucleotides, called a *codon*. For example, the codon GCT codes for an amino acid called alanine, whereas TTT codes for another amino acid, called phenylalanine. Because the DNA of our chromosomes is faithfully copied each time a cell divides, daughter cells inherit all the information stored in the DNA of the mother cell, including the sequences of all the proteins.

The DNA of our chromosomes functions a bit like a memory in a computer. When a cell needs to make a protein, the gene that codes for that protein is copied into another chain of beads called a *messenger RNA*, or *mRNA* for short. The mRNA is a copy of the gene DNA in a slightly different chemical form, called RNA. Being a faithful copy of the gene DNA, the mRNA contains the list of the amino acids and their order

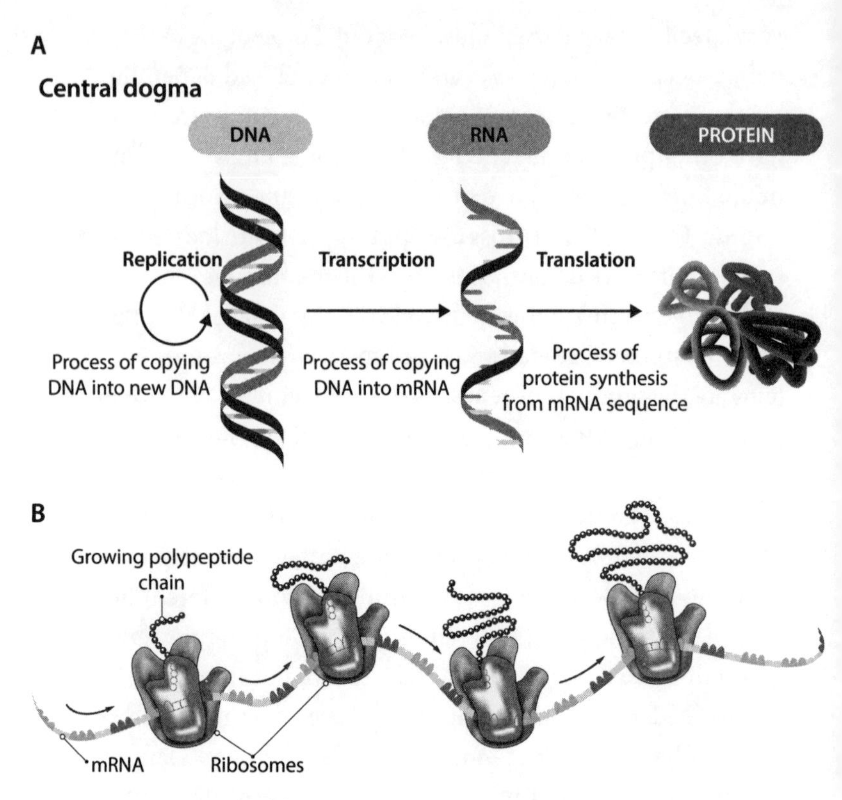

FIGURE 2.3. The mechanism of protein synthesis. (A) Pathway from DNA to messenger RNA to protein. (B) Ribosomes produce a protein by translation of its messenger RNA. A: Modified from BigBearCamera/Shutterstock; B: modified from EreborMountain/Shutterstock

in the protein. It functions like a program for a machine that makes proteins. This micromachine is called a *ribosome*. The mRNA is physically threaded through the ribosome. Free amino acids, the beads of twenty different colors, are available around the ribosome. Each time a three-letter codon on the mRNA passes through the ribosome, the ribosome selects the corresponding amino acid and links it to the nascent protein chain. It transforms the list of amino acids, the program, into a chain of amino acids linked to one another, the product.

Thanks to techniques such as X-ray diffraction and cryo-electron microscopy, the structure of ribosomes is known in detail, and the way they function is well understood. A ribosome is a complex assembly of three special RNA molecules and close to a hundred different proteins. This assembly is very large, making ribosomes visible as dots inside the cell under the electron microscope. Cells contain a very large number of ribosomes.

You may wonder how ribosomes are made by a new cell, since they are made of protein, and it is the ribosome that makes proteins. The answer is that there are always ribosomes already around. When a cell divides, each daughter cell inherits its part, usually half, of the cytoplasm of the mother cell, and therefore half of its ribosomes. They will be needed to make the proteins of the new cell, including those of its new ribosomes. Even the very first cell of any organism, the fertilized egg, contains ribosomes because the cytoplasm of the egg contains ribosomes.

Protein Folding

The chain of amino acids that comes out of the ribosome during protein synthesis is linear and flexible. With exceptions that we will discuss later, the chain needs to fold in a very specific way to make a mature, functional protein. This folded form is called the "native" form of the protein, because it is the form that the protein has when it is extracted carefully from a tissue or fluid. Figure 2.1 shows the native form of a protein called T4 lysozyme, which happens to be an enzyme. As already mentioned, the shapes of native proteins are extremely diverse, a reflection of the great diversity of amino acid sequences and chain lengths. So how does the linear chain of amino acids fold to form the native protein with its alpha helices, beta sheets, and many loops?

The process is extremely complex and is the object of much research. Because it is at the center of the problem of prions, we will summarize some essential aspects of protein folding in this chapter. The chain folds because the chemical structure of each amino acid along the chain allows it to stick, or not, to other amino acids in the same chain. We have already discussed how this can create alpha helices and beta sheets. It can also create all sorts of bonds between amino acids in the loops that connect the alpha helices and beta sheets.

To appreciate the complexity of the process, let's take the hypothetical example of a protein that is a hundred amino acids long. This is a chain of a hundred amino acids, each of which can have any of twenty different structures and chemical properties, all arranged in a precise order. Let's consider what may happen to a given amino acid in the chain, say amino acid number 25. Whether it will stick to another one in the chain depends on the chemical structure of number 25 and that of the other amino acid. Among the twenty different amino acids, some pairs will bind strongly, others weakly, some not at all. Number 25 may interact—strongly, weakly, or not at all—with any of the ninety-nine other amino acids in the chain. Any interaction will depend on many factors, including the distance between number 25 and the other amino acid along the chain. Furthermore, in its attempt at binding to another amino acid, number 25 is competing with all the other ninety-nine amino acids in the chain, which are all also trying to find a partner. The number of possible interactions during the folding of our hypothetical protein of a hundred amino acids is astronomical. Until recently, there was no computer program, and no computer powerful enough, to predict the folding of a protein from the sequence of its amino acids. It was just too complicated.

This all changed dramatically in 2021, when two groups of scientists published programs, named respectively AlphaFold and RoseTTAfold, that predict the folding of a protein from its amino acid sequence. AlphaFold was developed by DeepMind Technologies, an artificial intelligence start-up acquired by Google in 2014. RoseTTAfold came from a collaboration between academic laboratories led by the University of Washington. Both methods use machine learning, artificial intelligence, neural networks, and very powerful computers.

Before AlphaFold and RoseTTAfold became available, scientists had determined the structure of around 150,000 proteins using the slow, demanding techniques of X-ray crystallography and cryo-electron microscopy. AlphaFold and RoseTTAfold used artificial intelligence and machine learning to extract from these experimental results rules that link protein sequence to protein structure. The pace at which new structures have been determined by these new techniques is astounding. Only one year after AlphaFold was published, an article came out that described the structure of all the proteins whose sequence was available in the most complete database.

Not just human proteins, but proteins of every animal, plant, and microorganism for which a sequence was available. Another impressive aspect of this research is that all the results have been made publicly available—there is no restriction to their access. A nice case of openness in scientific research.

Proteins fold inside cells, where thousands of different proteins are packed close to one another, creating the possibility for all kinds of interactions that could interfere with proper folding. The inside of the cell is so crowded with molecules, mainly proteins, that it is more like a gel, or egg white, than a liquid. To prevent such harmful interactions, cells have special structures made of proteins called *chaperones*. These micromachines have a cavity in which a nascent protein emerging from ribosome is protected from interactions with other cell components and can fold correctly. But proteins can still sometimes "misfold"—things go wrong, and they fold in a way that is different from their "native" folded form, the one normally found in cells. Misfolded proteins can be toxic, and they are also a waste of resources, but cells are equipped to deal with them. Cells harbor special structures, also made of proteins of the chaperone family, that recognize misfolded proteins and either unfold and refold them, or send them to other structures, called *proteasomes*, where they are degraded to recycle the amino acids. Altogether, the various micromachines, cell structures, and reactions that deal with misfolded proteins constitute what is called *protein quality control*. There is evidence that in higher organisms, including humans, protein quality control declines with age. This decline may be one reason why aging is the main risk factor for neurodegenerative diseases associated with protein misfolding, such as Alzheimer's and Parkinson's disease.

Intrinsically Disordered Proteins and Prion Proteins

Some proteins do not fold spontaneously, even with the help of chaperones. They are called *intrinsically disordered proteins.* In most cases only a portion of the protein, what is called a *domain,* remains unfolded—intrinsically disordered—and the rest of the protein folds normally. To fold properly and acquire a function, this domain must bind to a partner, a specific cell component that in most cases is another protein. This mechanism is fittingly dubbed *binding-induced folding.*

Intrinsically disordered domains do not just hang there passively. We should think of them as attempting to fold in myriad possible shapes successively, each one lasting only somewhere between a micro- and a millisecond—a millionth and a thousandth of a second—a very short length of time, because they are unstable. They fold and unfold constantly in many different, unstable shapes until they encounter their partner. This encounter stabilizes one particular shape among all the unstable ones. That shape becomes the native folding of the protein domain. Proteins with intrinsically disordered domains are common. Scientists are now studying some that may have more than one partner, each partner stabilizing the protein in a different shape. This may be one of the various mechanisms that regulate the activity of proteins, changing their function according to the partner to which they bind.

Prion proteins are intrinsically disordered, or at least have an intrinsically disordered domain. They are a special class of intrinsically disordered proteins, having the property of being *self-templating.* This means that the protein can impose its own folding onto another copy of the same protein—that it can serve as a template for the folding of another copy of itself. How this is done is not entirely clear at present. The most likely mechanism, one that is consistent with experimental results and with present models of protein folding, is shown in figure 2.4A.[2]

Let's take the scrapie agent, PrP, as an example. The protein has a long intrinsically disordered domain. About one-quarter of the chain, at one extremity, is disordered. PrP is a normal neuron protein, whose function in the life of neurons is not well known. It must have a partner that correctly folds its intrinsically disordered domain and makes it perform its function. However, for PrP there is an unfortunate alternative to this normal behavior.

A neuron contains many copies of PrP molecules, with their intrinsically disordered domains folding and unfolding all the time until they find a partner. Among these many identical proteins with their myriad unstable, short-lived shapes, lets imagine that two of them happen to have the same unstable shape at the same time, that they bump into each other and bind, and that this encounter stabilizes the intrinsically disordered domain of both into that same previously unstable shape. In other words, in this case the partner that stabilizes the intrinsically disordered domain of PrP is another copy of PrP with its intrinsically disordered domain in the same shape—another copy of itself. Now, we have a stable pair of

2. This is an unpublished model proposed by Ronald Melki. It is similar to the *template-mediated selection* model proposed by others. Besides Ronald Melki, I have discussed this model with other experts in protein folding, including Daniel F. Jarosz from Stanford University.

PrP proteins with a new shape. Because it is stable, other copies of PrP with that same shape will encounter the pair, bind to it, and become stabilized. This will cause a buildup of proteins of the same shape all stuck to one another, a bit like a pile of soup dishes in the kitchen cabinet, which becomes taller as more dishes are added. Eventually this buildup creates a fibril large enough to be visible under the electron microscope. And strange, long fibrils were indeed observed a long time ago by scientists studying the brains of scrapie animals using electron microscopy. These fibrils can now be purified from the brains of infected animals (figure 2.4B). Each one comprises a very large stack of the PrP protein in its prion shape (figure 2.4C).

Now let's imagine that the fibrils of PrP get broken into smaller pieces. These pieces will act as seeds. They will trap more and more PrP molecules with the PrP prion shape and start building new fibrils in the neuron. The system amplifies itself. Some pieces may get out of the neuron and enter other neurons. When this happens, PrP fibrils start building up in these new neurons. The process can go on and on, and the fibrils will spread among the neuron population like an infection. PrP will behave as an infectious protein, a prion.

In chapter 1 we mentioned the possibility that the PrP prion could get "adapted" to a new host; for example, the adaptation of the sheep PrP prion to the laboratory mouse. The sequence of amino acids of PrP is nearly the same in all mammals, but only *nearly*. There are mutations specific to each species. Some of these mutations change one or a few amino acids in the protein. From the mechanism of self-templating and the structure of fibrils shown in figure 2.4, we might expect that the binding of a PrP molecule with a mutation to a PrP prion

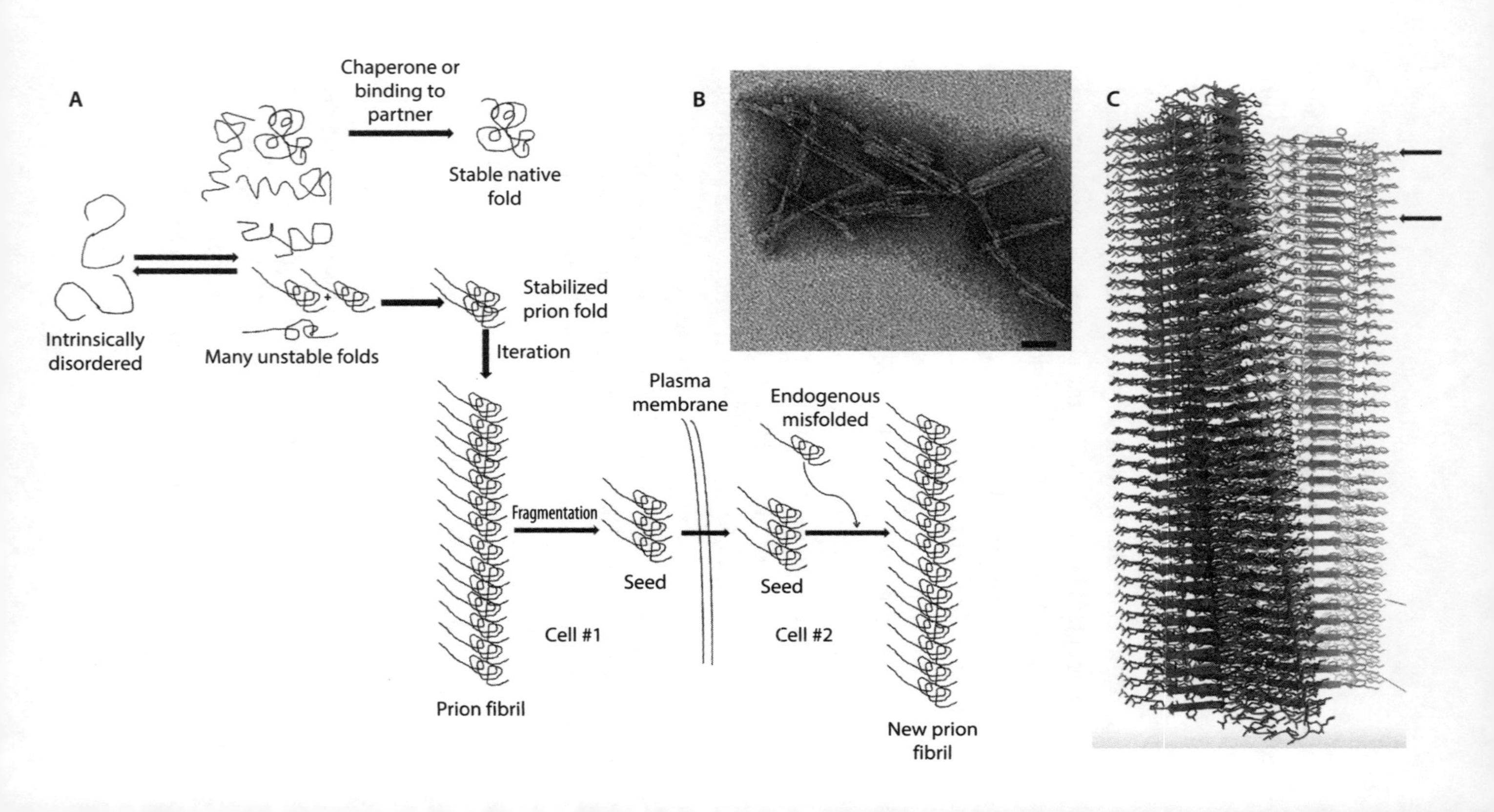
A
Chaperone or binding to partner
Stable native fold
Intrinsically disordered
Many unstable folds
Stabilized prion fold
Iteration
Prion fibril
Fragmentation
Seed
Cell #1
Plasma membrane
Seed
Endogenous misfolded
Cell #2
New prion fibril
B
C

FIGURE 2.4. (A) A model for the formation of a prion protein. An intrinsically disordered protein folds and unfolds in many unstable shapes. Binding to a partner stabilizes it into its native shape. But, as an alternative, two short-lived unstable shapes may bind to each other and form a stable pair. This original pair traps more short-lived unstable proteins with the same shape, forming a long fibril. Fragmentation of the fibril produces a seed. The seed can enter another cell and initiate the formation of a new fibril. The process repeats itself from cell to cell, spreading the prion through the organism like an infection. (B) Scrapie fibrils purified from the brains of mice, as seen under the electron microscope. The scale bar is 50 nm (a nanometer is a millionth of a millimeter). (C) Reconstruction of the structure of an elemental scrapie fibril. The fibrils seen under the electron microscope, as in B, are made of several such elemental fibrils sticking to one another. The structure was determined by cryo-electron microscopy. The stacking of 35 identical molecules of PrP is clearly visible. The arrows on the right point to the extremities of two PrP molecules within the stack. B and C: From "Cryo-EM of Prion Strains from the Same Genotype of Host Identifies Conformational Determinants" by Forrest Hoyt et al. (*PLOS Pathogens* 18, no. 11 [2022]: e1010947); Creative Commons CC0 public domain dedication seed from the original host, and therefore without mutation, may not be possible. The change in amino acid sequence might interfere with the binding to the PrP seed and the stacking of the molecules to form a fibril. This is why in general PrP prions have a narrow host range. A prion seed from sheep will have difficulty binding PrP in mice. However, depending on the species and their various mutations, this may be only a difficulty. Rarely, with low probability, binding may occur, and a sheep PrP prion seed binds and stabilizes a mouse PrP into a prion shape. Once this happens it triggers the formation of a mouse PrP prion fibril and the propagation of the new prion. The low probability of such an event is why adaptation to a new species requires repeated serial inoculations and a long time.

You may ask: What about protein quality control, which is there to refold or degrade misfolded proteins? It should prevent self-templating. This is a good question. Quality control certainly interferes with the formation of prion proteins. Remember that scrapie, Creutzfeldt-Jakob disease, and others prion diseases are rare and slow diseases. It takes years for the lesions to spread in the brain to the point of impairing its function. During this time there is competition between the buildup of prion proteins and quality control. However, with time, quality control may be overwhelmed so that the disease shows itself with its symptoms.

There is also a sinister twist in the competition between prion protein buildup and quality control. To be infectious, the prion fibrils need to be broken into small pieces to form seeds that are small enough to spread from cell to cell. Ironically, this breaking down is due to protein quality control. Quality control is performed by a complex collection of different proteins and micromachines. One of them breaks down large assemblies of misfolded proteins so that they can then be degraded. But this process might be swamped by the abundance of material and may produce seeds from fibrils of stacked prion proteins.

Why is the PrP prion toxic for neurons? This is another part of the story for which we do not have a definitive answer. The normal PrP molecule is found at the surface of the neuron, attached to the outer side of the neuron plasma membrane. The prion form of PrP—the PrP fibrils—accumulates outside the cell. It appears that the PrP prion fibrils may interact with the normal PrP and make the cell membrane leaky, or at least permeable to certain toxic molecules. Another possible mechanism of toxicity, which applies to all the prion proteins responsible for diseases that will be considered later in this book, is simply that the accumulation of the protein in prion fibrils deprives the neuron of the normal form of the protein and its function. This is toxicity by starvation—by depriving the cell of a necessary component. A lot more work needs to be done to gain a clear picture of the events that lead to the death of neurons in spongiform encephalopathies and other prion diseases.

The idea that proteins can behave as infectious agents through the mechanism that we just described was, at first, hard to accept for microbiologists. Some—we mentioned Laura Manuelidis from Yale University—are still unconvinced. But there is always a tipping point with revolutionary concepts in science. Ideas that were questionable all a sudden become general knowledge, and the story of the discovery tends to be forgotten.

However, the story of prion proteins is still ongoing. This book will now deal with new developments in the field. We will show that prion proteins other than PrP are associated with common diseases such as Alzheimer's and Parkinson's, and even type 2 diabetes. We will also describe how prion proteins associated with disease may be the exception. Indeed, there are good prions! Prion proteins that perform important functions in the cell. Prion proteins appeared very early during evolution and are present in primitive organisms such as yeast and mushrooms. They were retained during evolution because of their useful functions. Study of the origin and evolution of prion proteins is still a new field of biology, one that, as you might expect, is not devoid of controversy. However, neuroscientists have discovered a prion protein that is important for the acquisition and recall of memory. Immunologists have found prion proteins that are essential for defense against viruses. The list of good prions will undoubtedly get longer. It is safe to say that we still have a lot to learn about good prion proteins *and* prion proteins that cause disease. These are exciting developments, which may lead to new treatments for some dreadful diseases, as well as broadening our understanding of biology in general.

To Recap

Proteins are very diverse in size, shape, and function. They are like chains made of beads of twenty different colors. The "beads" are small molecules called *amino acids*. Proteins acquire their shape and function after the chain of amino acids folds. The structure of folded proteins is made up of *alpha helices* and *beta strands* connected by *loops*.

The detailed structure of proteins, down to the position of every atom, can be determined by several methods, including

X-ray diffraction of protein crystals and cryo-electron microscopy. In 2021, two groups of scientists published computer programs that predict the folded structure of proteins from the sequence of their amino acids. This was a major achievement.

Proteins are made by cells. Cells are separated from their environment by a *plasma membrane*. There are other membranes inside cells that separate various organelles, including the nucleus with its nuclear membrane. The amino acid sequence of a protein is encoded in DNA, in the gene that codes for that protein. The gene's DNA is transcribed into RNA, which is decoded—translated—into the chain of amino acids by micromachines called *ribosomes*.

Protein folding—the folding of the amino acid chain to give the protein its final shape—is a very complex phenomenon. Misfolding can happen and may be toxic for the cell. *Protein quality control* takes care of "misfolded" proteins.

Some proteins are called *intrinsically disordered*. To fold into a final stable shape, they must bind to a partner, which is usually another protein. Prion proteins are a special case of intrinsically disordered proteins. They can fold into a normal functional shape by binding to a partner. They can also fold into a prion shape that is *self-templating*. This means that the prion shape can cause other copies of the same protein to fold into an identical prion shape. These folded prion-shape proteins bind to one another and eventually form a long fibril. Fragments of the fibril that are broken off act as *seeds* that initiate more prion folding of the same protein and the buildup of more fibrils. If seeds enter new cells, they fold the protein in these new cells into prions. The prion protein thus spreads like an infectious agent.

CHAPTER 3

The Brain, Our Most Fascinating Organ

It is impossible not to be fascinated by what our brain can do. While I am focusing on writing this new chapter it keeps track of where I am (at my desk in Palo Alto), of the time of day (early afternoon), of the temperature around me (an uncomfortable and unusual heat wave), and on the reason I am imposing this on myself (I promised myself not to diddle any longer and start writing now). Besides all this, of which I am conscious, my brain constantly monitors what is going on in my body and performs all sorts of adjustments to keep it functioning properly. I am not conscious of all these adjustments. My brain does all this without bothering me with such boring housekeeping questions.

Brains, from those of the simplest animals all the way up to our own, have evolved to sense both the outside, the environment, and the inside, the interior of the organism, and to integrate the information in order to "decide" a course of action. For most of the animal kingdom the decisions have to do with staying away from predators, finding food, and finding a mate to produce offspring. To this basic machine, many millions of years of evolution have added layer after layer of complexity and refinement, which have progressively increased its performance to the point where you can read this book and hopefully find it interesting, and I can enjoy the beauty, and the complexity, of J. S. Bach's Goldberg variations. At present, the human brain is definitively the most elaborate product of the evolution of life on the planet.

Neurons Are Functional Units in the Brain[1]

The basic functional unit of a brain is a cell called a *neuron*. Despite the millions of years of evolution that separate us from snails, flies, or sea urchins, our neurons function in essentially the same way as those of these primitive animals. This tells us that the appearance of neurons during evolution must have been a revolution with immense consequences for the story of life. Without it I would not be writing this book, and you would not be reading it. It also has a profound practical consequence for neuroscientists. It means that we can study fundamental aspects of our brain using very simple—and therefore easy to experiment with—organisms, such as a tiny worm with exactly 302 neurons (*Caenorhabditis elegans*), a sea slug (*Aplysia*) whose neurons are so large that some of them can be seen with the naked eye, and the fruit fly *Drosophila* that is a favorite organism of geneticists.

1. An excellent introduction to basic concepts of neuroscience is to be found in a wonderful book by Eric Kandel called *In Search of Memory* (London: W.W. Norton, 2006).

Everybody knows about neurons and the fact that we supposedly—this is now challenged by recent results—lose them as we age. But there are many other types of cells in our brain. *Astrocytes* and *oligodendrocytes* are also very important cells. Astrocytes perform a host of different functions that we will not discuss here. Oligodendrocytes make the *myelin* that surrounds many axons and increases the speed at which nerve impulses travel along them. *Microglial cells* are also very important. They sample their environment and detect pathogens, as well as other structures that should not be present in the brain, and they initiate an immune response against them. They play an important role in Alzheimer's disease and several other neurodegenerative diseases. Astrocytes, oligodendrocytes, and microglial cells, as a group, are referred to as *glial cells*.

Cells and Neurons

By the middle of the nineteenth century, scientists had concluded that plants and animals are made of cells—small, separate structures packed next to one another. However, they thought that the brain was an exception. Under the microscope, the brain appeared to be made of small bodies that looked like cells but with a very large number of tangled, intermingled extensions, which did not allow one to trace boundaries between those cells. The view at the time was that these extensions formed a web of continuous connections between the cell bodies. This was the *reticulum* theory of the brain, and it was very popular. One of its main advocates was Camillo Golgi, a famous Italian anatomist.

In the 1880s, the Spanish neuroanatomist Santiago Ramón y Cajal examined the brain using a staining technique developed, ironically, by Golgi. The technique stains only a very small number of neurons in a brain sample. But the neurons that are stained are stained in their entirety. With Golgi's technique, Cajal observed, and illustrated in wonderfully detailed and artistic drawings, that neurons are individual cells with many long slender extensions, but that they are separate from one another (figure 3.1). He also described contact points between these cells and did not see any evidence for continuity. He called these contacts *synapses*. The neurons were distinct, separate cells. This observation was the founding event of modern neuroscience. The relationship between Golgi and Cajal was tumultuous, to say the least. When they both received the Nobel Prize in 1906, Golgi for developing the technique and Cajal for showing that neurons are separate entities, Golgi, to the embarrassment of the Nobel committee, still believed in his reticulum theory.[2]

The role of Ramón y Cajal in neuroscience cannot be overstated. From pure observations under the microscope, he inferred the fundamental principles that govern brain functioning. He concluded that neurons are the functional units of the nervous system. He described how neurons are made of three continuous essential parts: the *cell body*, also called the *soma*; its numerous thin extensions, the *dendrites*; and a single extension that looked different from dendrites, which he called the *axon*.

2. Ramón y Cajal wrote an autobiography, which makes for very good reading: *Recollections of My Life*, which has been translated by E. Horne Craigie and Juan Cano (Cambridge, MA: MIT Press, 1989).

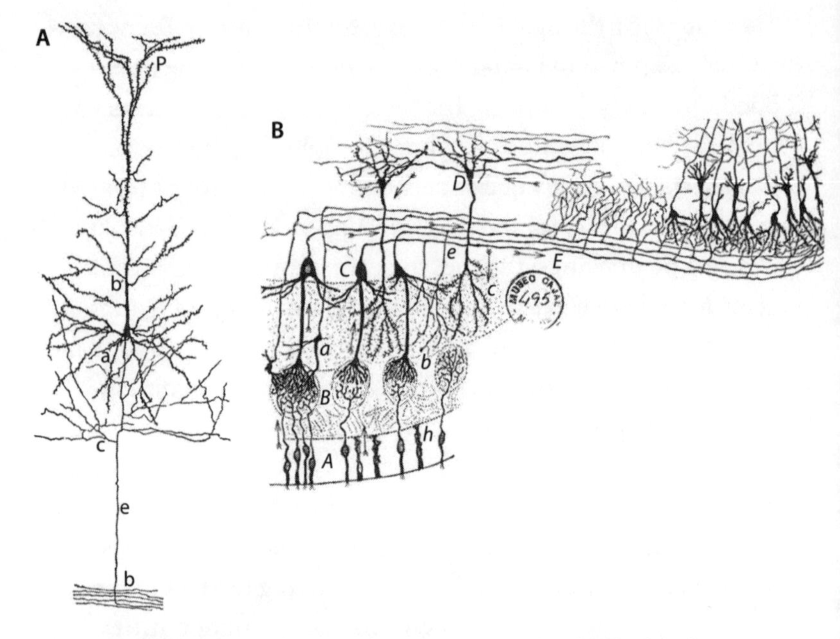

FIGURE 3.1. Two drawings by Ramón y Cajal published in the first volume of his book *Textura del systema nervosio del hombre y de los vertebrados*, published in Madrid in 1899. (A) A neuron in the cortex of the rabbit brain. The axon is labeled *e*; *a* points to several dendrites; *c* indicates a bifurcation of the axon. (B) The network of neurons for olfaction. The layer of olfactory neurons in the nose, at the bottom, is labeled *A*. The arrows indicate the direction of the nerve impulse from the olfactory neurons in the nose to the brain, as correctly intuited by Ramón y Cajal. From *Textura del systema nervosio del hombre y de los vertebrados* by Santiago Ramón y Cajal (Moya, Madrid, 1899)

He intuited that information flowed in only one direction, from the dendrites to the cell body to the axon, and from there to other neurons through contacts between axons and dendrites, contacts that he called *synapses*. He described these contacts and the small gap, the *synaptic cleft*, that separates the two neurons. He went further and described how neurons, through synaptic contacts, make specific circuits that extend across the brain. From observations with his microscope, he concluded that information must flow in one direction in these circuits, from the axon of one neuron to the dendrites of a different neuron, and that the circuits must be the substratum of specific brain functions, such as vision or smell (figure 3.1 B).

However, after Cajal had firmly established the notion that the neuron is the fundamental unit of the nervous system, a host of questions still needed to be answered in the burgeoning field of neuroscience. Cajal's work consisted purely of observations through the microscope, and in that way it wasn't possible to answer questions such as what the nature is of the information flowing down axons, or how information is transferred from one neuron to another at synapses. Answering these questions required experiments.

Neurons Fire Action Potentials

Studying the nature of the information flux became the field of electrophysiology. At the end of the eighteenth century, Luigi Galvani had shown that nerves connecting the spinal cord to muscles carry electrical currents. In the nineteenth century, Hermann von Helmholtz discovered that this electrical current is very different from that going through a metal wire. For one thing it is much slower. However, although the potential is very weak, a mere 70 millivolts, the current does not decrease as it travels, even over long distances. We now know that this is because the current travels like a pulse, being constantly regenerated by a movement of sodium and potassium ions across the membrane of the axon (figure 3.2). The current

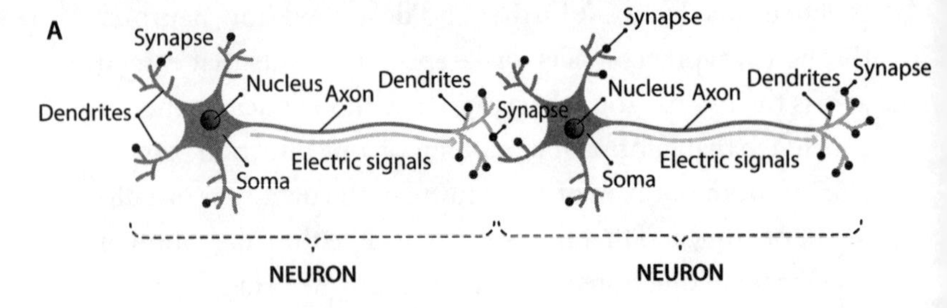

FIGURE 3.2. (A) Diagram of two neurons connected by a synapse between the axon of the neuron on the left and a dendrite of the neuron on the right. (B) A neuron at rest receiving a stimulus at a synapse. The interior of the axon is negative relative to the exterior. (C) The neuron fires an action potential down its axon. In the area of the action potential, the electrical charges are reversed by movement of ions across the axon membrane. (D) Potassium ions move inside the axon and repolarize the membrane, but this time with more potassium ions outside than inside. That makes this part of the axon *refractory* to depolarization, meaning the action potential can only move down the axon. A: VectorMine/Shutterstock; B: udaix/Shutterstock

travels along the nerve, from one regeneration to the next, in a kind of domino effect.

Ions are atoms with an electrical charge. The charge can be positive if it is due to the loss of an electron, which is the case for sodium and potassium ions, or negative if the atom gains an electron—for example, a chloride ion. In the brain, ions travel through what are called *ion channels,* complex assemblies of several proteins lying across the axon's cytoplasmic membrane, with one face on the outside and one face on the inside. The proteins that make up an ion channel lie very close to one another, but there is a tiny gap, a pore, in between them

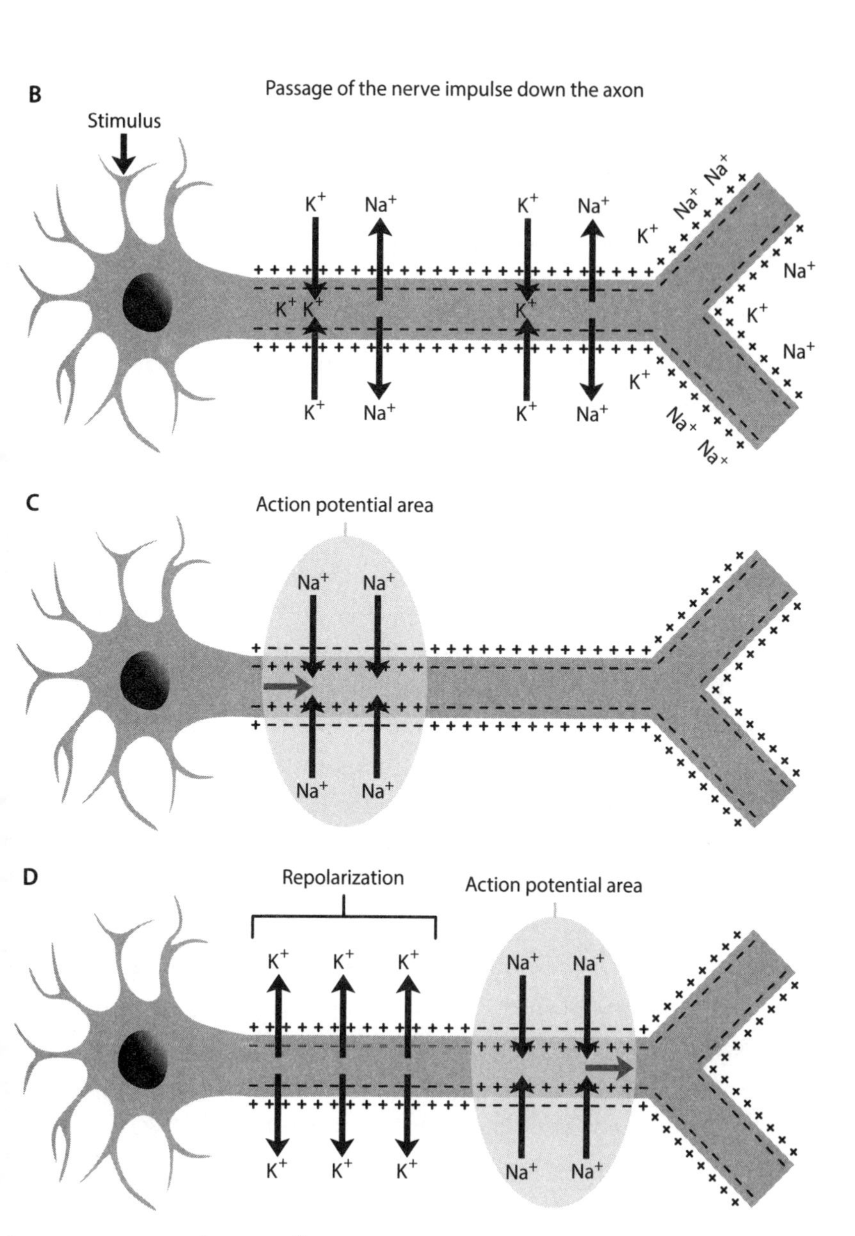

FIGURE 3.2. (*continued*)

that can let ions pass through. Channels are specific for one sort of ion only. There are sodium channels, potassium channels, calcium channels, and chloride channels. Many of these channels are not passive structures: depending on the conditions they can be open or closed.

Both sodium and potassium ions carry a positive electrical charge. In the brain, there are more sodium ions outside neurons than inside, and more potassium ions inside than outside. At rest, some potassium ions diffuse out of the axon through the potassium channels, which stay open, since the concentration of potassium is higher inside than outside. The sodium ions cannot diffuse into the axon because the potassium channels let only potassium to go through and because the sodium channels are closed. This creates an excess positive charge on the outer surface of the axon membrane, so the outside becomes positively charged and the inside negatively charged. The electrical potential between inside and outside is 70 millivolts (a millivolt is a thousandth of a volt).

To understand how this relates to information flux in the brain, we need to consider special channels called *voltage-gated channels*. These are also ion specific, allowing either sodium or potassium to cross the membrane. At rest they are closed and ions cannot pass through them. But when a stimulus in the form of an electrical current reaches a voltage-gated sodium channel, the channel opens briefly, letting sodium ions, with their positive charges, enter the axon. The resulting increase in positive charge inside the axon causes the difference in electrical potential between the outside and the inside of the axon membrane, the voltage, to disappear, or even overshoot in the other direction. This has two consequences. First, it closes the sodium channels that had been opened by the stimulation. Second, it opens voltage-gated potassium channels, which, by letting potassium ions exit the axon, restore the original distribution of electric charges on either side of the membrane and return the difference in potential to its resting level of 70 millivolts, with more positive charges on the outside than on the inside. The whole event lasts only a millisecond and is called an *action potential*. Note that now the difference in potential is due to an excess of potassium ions on the outside and a paucity of sodium ions on the inside, the opposite to the previous resting situation. The distribution of ions across the membrane will return to normal, with more sodium outside and more potassium inside, after a short delay called the *refractory period*. This return to normal is achieved by special ion channels that can use energy to "pump" potassium ions across the membrane toward the inside of the axon. It is called a "refractory" period because a new action potential cannot happen while the distribution of ions is reversed.

The action potential creates an electrical current, an excess of electrons, which diffuses along the axon, reaching more voltage-gated sodium channels further down the axon, and the process repeats itself all the way along the axon in a domino-like effect. But there is no reason why electrons should diffuse in only one direction, so why does the current always move toward the end of the axon? This is because of the refractory period mentioned above. Voltage-gated channels cannot function again for a brief period after an action potential. They are *refractory* to an incoming electrical current. So the current can trigger a new action potential in only one direction: away from the previously active voltage-gated channel, therefore away from the cell body and toward the end of the axon.

The current that travels along the axon, one voltage-gated channel at a time, is not very fast. Its low speed is sufficient for some functions in the body, such as transmission of the sensation of pain from the skin to the brain. But for other functions, such as movement of limb muscles, for which axons can be more than a meter long in large mammals, including humans, transmission needs to be faster. This is achieved by something called *myelin*.

Myelin is made of many layers of membrane wrapped around the axon. These membranes are an extension of the plasma membrane of specialized cells: oligodendrocytes in the brain and spinal cord, Schwann cells in peripheral nerves. Myelin works like an insulator around a copper wire, and prevents the electrical current from leaking into the surrounding tissue. Under the myelin the current, the electrons, spreads by simple diffusion at high velocity. However, an axon is not like a copper wire. The current that diffuses down the axon beneath the myelin insulator decays rather rapidly. This problem is taken care of by what are called *nodes of Ranvier*, named after Louis-Antoine Ranvier, the neuroanatomist who first described them in the second half of the nineteenth century. Nodes of Ranvier are evenly distributed along the axon. They are short spaces without myelin and with voltage-gated sodium and potassium channels. When the weakened electrical current reaches a node of Ranvier, it generates a new strong action potential, which quickly diffuses to the next node of Ranvier, and so on. The action potential travels down the nerve by jumping from one node to the next. In diseases such as multiple sclerosis, myelin breaks down. Conduction becomes slow, affecting movement in the arms and legs. Eventually, axons without a myelin sheath suffer and degenerate.

Neurons Communicate with Each Other at Synapses

Neurons communicate with each other through synapses, the points of contact between two neurons. At the synapse, the membrane of an axon terminal of the upstream (*presynaptic*) neuron comes into close contact with the membrane of a dendrite, or the soma, of the downstream (*postsynaptic*) neuron. Synapses are visible under the electron microscope (figure 3.3). The dendrite side of the synapse appears as a globular structure with a short stem and contains many different proteins (figure 3.3C). Both sides of the synapse also contain a collection of messenger RNAs coding for synaptic proteins.

Some synapses are *electrical synapses*. The electrical current of the first neuron crosses the synapse and triggers an action potential in the second neuron. These synapses are fast and are used, for example, in reflex responses.

Most synapses are not electrical but *chemical synapses*. At the axon terminal, substances called *neurotransmitters* are stored in small *synaptic vesicles*, which are visible under the electron microscope (see figure 3.3B). When an action potential reaches the axon terminal, synaptic vesicles fuse with the axon membrane and release their neurotransmitter into the *synaptic cleft*, the narrow space that separates the two neurons. The neurotransmitter diffuses across the synaptic cleft and binds to a receptor on the membrane of the downstream neuron. The

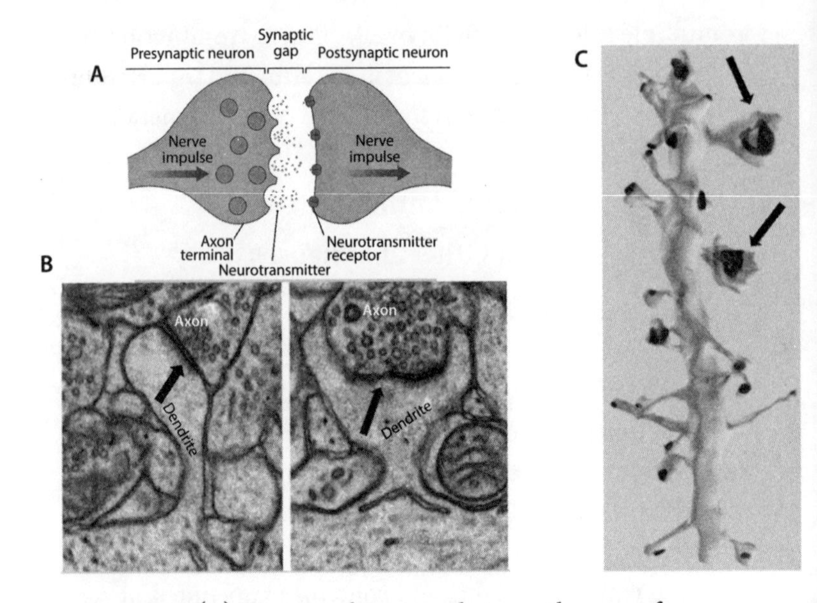

FIGURE 3.3. (A) Diagram of a synapse between the axon of a presynaptic neuron and a dendrite of a postsynaptic neuron. (B) Synapses as seen under the electron microscope. Note the synaptic vesicles, which contain the neurotransmitter, in the axon terminals. The arrows point to the synaptic cleft and its thin gap. (C) Reconstruction of a dendrite observed in serial sections with an electron microscope. The dendrite has many synapses, most of them small. Two (arrows) have been remodeled and enlarged. A: Modified from Ali DM/Shutterstock. B: From "Ultrastructure of Synapses in the Mammalian Brain" by K. M. Harris and R. J. Weinberg (*Cold Spring Harbor Perspectives in Biology* 4 [2012]: a005587); reproduced with permission. C: Unpublished; reproduced with permission from J. Spacek

brain uses many different neurotransmitters, each with its corresponding specific receptor. As a rule, a given synapse uses only one kind of neurotransmitter. Binding of a neurotransmitter to its receptor triggers an electrical current in the dendrite; this current is called a *graded potential* because its strength, measured in millivolts, depends on several factors and can vary.[3] If the strength of the graded potential reaches a certain threshold, it triggers an action potential in the dendrite, which will propagate all the way to the soma of the neuron.[4] Electrophysiologists say that the dendrite "spiked." The cell body sums up action potentials coming from several synapses and dendrites. The sum determines whether the neuron fires an action potential down its axon.[5]

Why should the brain use such complicated synapses when it is also equipped with the much simpler electrical synapses? Chemical synapses are important because they can modulate the intensity of the transmission of the electrical signal between two neurons. They could be compared to a dimmer switch that produces more or less light depending on the local circumstances. Synapses, dendrites, and neuron cell bodies are the essential components that make the brain a decision-making machine.

3. The strength depends, among other things, on the frequency with which action potentials reach the synapse. All action potentials have the same amplitude (70 millivolts) and the same duration, a millisecond (1/1000 of a second). As a rule, a single action potential is a very weak signal that does nothing to the synapse. On the contrary, a train of 100 action potentials within a second is an intense signal that will trigger the release of neurotransmitter.

4. For some time, it was believed that dendrites did not have voltage-gated ion channels as axons do, and that the graded potential diffused passively to the soma of the neuron. This was incorrect. It is now well established that action potentials propagate along dendrites.

5. This is an oversimplified description. Among other things, dendrites by themselves can sum action potentials from several synapses, thereby sending a strong electrical signal to the soma.

Synapses are not fixed structures. During the lifetime of an animal, or human, they can be weakened, even eliminated, or strengthened. After birth, while the young animal is acquiring new skills, synapses that are repeatedly stimulated undergo what is called *synaptic remodeling*. This is a complex process where new proteins are synthesized and incorporated into the synapse. This causes a visible enlargement of the synapse and often the creation of new synapses nearby connecting the same two neurons. Conversely, synapses that are not used are eliminated. We will see in chapter 7 that synaptic remodeling is essential for the acquisition and retrieval of long-term memory.

Neurons Come in Many Different Types

You should not conclude from the previous paragraphs that all neurons are identical and function in the same way. All neurons have a cell body, dendrites, and an axon and communicate with other neurons at synapses. But within this common frame there are many variations. The shapes of neurons vary depending on their position in the brain and their function, as illustrated in Cajal's drawing of the olfactory network of neurons, shown in figure 3.1B. The number and the degree of branching of dendrites can vary considerably. Axons may have a few or many branches. These differences in structure reflect functional and chemical differences. Although, with a few exceptions, each synapse uses only one neurotransmitter, more than one hundred different neurotransmitters have been identified. Among them, glutamate is a neurotransmitter used by many excitatory neurons, whereas one called GABA is widely used by *inhibitory neurons*, neurons that prevent the neurons they are connected to from firing action potentials. And shape and neurotransmitter are only two characteristics of neurons among many others. This cursory description gives you an idea of the large diversity of neuron types in a brain like ours.

Neurons Are Not the Only Kind of Brain Cells

Besides neurons, many other sorts of cells are present in the brain, in particular a group of three different cells collectively called *glial cells*. We have already talked about oligodendrocytes, the cells that make the myelin sheath that surrounds axons. We will later talk about microglial cells, which play important roles in the immunity of the brain and in the remodeling of synapses. We need to introduce the third kind, called *astrocytes*. They are numerous, ubiquitous, star-shaped cells. In the normal brain, among many other roles, they contribute to the correct functioning of synapses. Together with microglial cells, they also react to injuries. They become activated, and then they enlarge and proliferate. If the injury is an infection, they contribute to the immune response, and to the inflammation, in the fight against the pathogen.

Studying Neuronal Activity in the Whole Organism

Between the functioning of single neurons and synapses to that of a complex brain, there is of course an enormous gap that needs to be filled. Despite the difficulties, remarkable advances have been made in this aspect of neuroscience, thanks in great part to recently developed techniques. We will review some of them because they are relevant to the study of memory, a central issue in this book, and because they may give the reader a more concrete view of what neuroscientists do in their laboratories. These techniques take advantage of advances in the genetic manipulations of cells and organisms. They get more sophisticated and more powerful all the time. They also benefit considerably from recent developments in imaging the activity of single cells in tissues and organs.

Detecting Neurons Firing in the Brain in Live Animals

Neurologists and neuroscientists can detect the activity of neurons inside the brains of patients or laboratory animals, using techniques such as electroencephalography (EEG) or functional magnetic resonance imaging (fMRI). These techniques have been around for some time but are still extremely useful. However, their spatial resolution is not very good; at best of the order of 1 millimeter for fMRI. There might be fifty thousand to a hundred thousand neurons doing different things in such a space. How can we get a closer look at individual neurons while they are involved in specific tasks, such as acquiring the memory of a recent event?

Firing of action potentials is always associated with an increase in the concentration of calcium ions in the cytoplasm of the neurons. Over the years, molecular biologists have engineered proteins that emit light of a given color when they are activated by calcium ions. Therefore, neuroscientists thought of introducing these light-emitting proteins into the neurons that they wanted to study, to detect which one is firing at any given moment. How could that be done?

Proteins are the products of genes. Therefore, a way to make neurons produce a light-emitting protein is to introduce the gene for that protein into the neurons' DNA—this is called making the neurons *transgenic* for the light-emitting protein. With mice, this can be done in two ways. One is to make the whole mouse transgenic for the gene. The DNA that codes for the gene is introduced in the nucleus of a fertilized mouse egg, which is then implanted into the uterus of a foster mother. The foreign DNA gets integrated at random in the mouse DNA. The pup carries the foreign DNA in all its cells. But, for the gene to be expressed and the light-emitting protein to be synthesized, the gene needs to be preceded in the DNA by a sequence called a *promoter*. The promoter controls when and in which kind of cell the gene will be expressed and the protein synthesized. By a judicious choice of promoter, it is possible to express the transgene only in neurons, and even only in the category of neuron that one wishes to study.

The other method consists in introducing the gene for a light-emitting protein into a viral vector. A viral vector is a virus that has been genetically disabled. It does not multiply in or kill the cell it infects, or cause any other kind of lesion leading to a disease, but it carries the gene coding for the light-emitting protein in its genome. It can enter a neuron in the same way a normal virus would and deliver the gene to the nucleus of the neuron. Here again a judicious choice of promoter ensures that the light-emitting protein is made only by the neurons of interest. The big difference from using transgenic mice is that the virus can be delivered precisely, by microinjection, to a group of neurons of interest. Only these neurons will be transgenic. Also, keeping lines of transgenic mice in the laboratory is cumbersome and expensive. With viral vectors one can use normal mice that are widely available.

How can scientists detect the light emitted by single neurons in the brain of a live mouse? This is an engineering problem that has been solved by miniaturization. There are now tiny microscopes with fiber optics, which can be inserted into the brain at a desired position, and which detect light-emitting neurons in real time while the animal is moving around or performing a task.

Making Neurons Fire with Light

Optogenetics is a revolutionary method in neuroscience that allows scientists to control the activity of neurons with light. The idea of using light to make neurons fire action potentials goes back several decades. At the time, the technical challenge was enormous. But spectacular progress has been made in recent years by several scientists. A decisive step was achieved in 2005 by Karl Deisseroth and Edward Boyden at Stanford University. They decided to use channelrhodopsin-2, a protein found in a microscopic alga. Channelrhodopsin-2 is a transmembrane ion channel. It inserts itself in the cell membrane, with one extremity on the outside and the other on the inside of the cell. It opens when briefly illuminated with blue light and lets sodium ions enter the cell. Deisseroth and Boyden made mouse neurons transgenic for the protein and found the conditions under which the algal protein produced an action potential when briefly illuminated with blue light from a laser.

This was a breakthrough, and it led to many modifications and refinements. There are now proteins that react to light of different colors and some that block firing instead of producing an action potential.

Optogenetics is now widely used in neuroscience. The genes for various light-sensitive ion channels are introduced into neurons of mice or other laboratory animals, either with transgenic techniques or using viral vectors, as described in the previous section. Fiber optics and lasers are used to send light pulses to neurons in free-moving laboratory mice. Miniature microscopes connected to computers record the firing of the neurons connected to the one stimulated by light. Experiments that could only be dreamed of by neuroscientists a decade ago are now performed routinely in many laboratories. In chapter 7, we will see some examples for the study of the mechanism of memory.

The Immunology of the Brain Is Different

The immune system is everywhere in the body. It fights infection and cancer by detecting the presence of foreign, abnormal products, mainly proteins from invading microbes, and by eliminating the invaders using a variety of "immune reactions." Things can go wrong, and the immune system may also contribute to chronic diseases, including neurodegenerative diseases such as multiple sclerosis, Alzheimer's, and Parkinson's.

For a long time, the question of whether immune reactions take place in the brain was controversial. It goes back to the early 1950s, when Sir Peter Medawar, then at University College London, was studying the role of the immune system in rejection of skin grafts. When skin from one mouse is grafted onto the skin, or other organs, of a different mouse, it is quickly rejected by an intense immune response, unless the mice are genetically identical. Medawar observed an exception to that rule. If the piece of skin was grafted in the brain, or in the eye, which is an extension of the brain, it survived for a long period of time. He interpreted this result as showing that the immune system has no access to what is going on inside the brain. This was the origin of the dogma of the "immune-privileged" nervous system. For a long time, the idea that the brain was not surveyed by the immune system had a strong influence on neuroscience and immunology.

The immune-privileged-brain dogma fitted nicely with the existence of the blood-brain barrier. Scientists knew that when some dyes, such as methylene blue, are injected into the blood stream of mammals, they quickly diffuse into all the organs and turn them blue. All organs, that is, except the brain and spinal cord—the central nervous system. There appeared to be a barrier that separated the blood from the central nervous system. We now know a lot about this barrier. Indeed, the capillary blood vessels in the brain are different from those of other organs. They do not let molecules simply diffuse from the blood into the organ. On the blood-vessel side, the cells that line the capillaries stick to one another owing to the presence of what are called *tight junctions*. On the brain side, specialized cells, astrocytes, line the entire barrier. This does not mean that nothing can enter the brain through this barrier, but what enters has been selected and transported through the barrier. The barrier acts as a screen that chooses what it will deliver to the other side, into the brain.

Another argument in favor of the immune-privileged brain was the absence in the brain of lymph vessels, lymphatics, which are present in most organs. They drain waste products and some cells from the organ to local lymph nodes, which are the hubs of the immune system. It is in lymph nodes that the immune response against a microbe—production of specific antibodies and specific T lymphocytes—is initiated. The brain tissue, with its neurons and glial cells, is devoid of lymphatics. However, it was recently discovered that the membranes that separate the brain and spinal cord from the surrounding bones, the *meninges*, are very rich in lymphatics and immune cells. The meninges, among other functions, act as an interface between the central nervous system and the immune system.

As time went on and more discoveries were made, the dogma of the immune-privileged brain began to fall apart. It is now clear that the brain and the immune system do interact. The brain is equipped with several mechanisms to detect invaders and to bring in an immune reaction against them, but these mechanisms are specific to the nervous system.

A detailed description of brain immunology is of course outside the scope of this book. However, one cell type, the microglial cell, must be introduced because it plays a central role in neurodegenerative diseases, including Alzheimer's disease. These cells are tiny and have many slender extensions. For a long time, nobody took much interest in them; it is only in the past few decades that we have learned how important they are. They constantly move around the brain and send their extensions in all directions to sample their environment. This has been demonstrated by beautiful in vivo imaging. They act as police, making sure that everything is all right in the neighborhood. If they detect an injury, such as an infection or a stroke, they react swiftly, migrate to the site of the injury, multiply, and engulf and degrade damaged tissues and try to clean up the mess. It is fascinating to think that inside our brains, in between neurons and other immobile cells, microglial cells are moving around, in all directions putting out and retracting extensions, looking a bit like the hands and fingers of a pianist practicing scales!

If microglial cells detect the presence of a foreign molecule, they initiate an immune response. They send signals that attract immune cells, which cross the blood-brain barrier and move toward the invader. This causes what is called *inflammation*: the secretion by the microglial cells and the immune cells of toxic substances aimed at fighting the invading pathogen. Unfortunately, the inflammation, if prolonged, can be very damaging to the surrounding cells, including neurons.

Before leaving microglial cells, we should mention that they also play essential roles in the healthy brain. They perform what is called *synapse pruning*. Neurons form large numbers of synapses with other neurons during the development of an embryo. Once the fetus and then the newborn starts using its brain, the synapses that are needed for basic tasks such as moving limbs are stabilized and reinforced. Other synapses, which are not needed or which might even scramble the communication between neurons, are eliminated, or pruned. Pruning is done by the microglial cells. Somehow microglial cells appear to be "aware" of synapse activity. The exact nature of the chemical signals exchanged between synapses and microglial cells is still unknown. In any event, synapses that need pruning are engulfed and digested by microglial cells. At present it is not clear if synaptic pruning by microglial cells continues after development, in the adult. Clearly, we have not reached the end of the story of microglial cells and their constantly moving extensions playing scales around our neurons!

To Recap

The human brain is the most elaborate product of life on our planet. Neurons are the functional units of the brain. Neurons function in the same fundamental way in all animals, from tiny worms to flies to humans.

In the late nineteenth century, Ramón y Cajal determined the basic organization of the brain solely from observations under the microscope. He showed that neurons have long extensions: several *dendrites* and one *axon*. He intuited that information flows through neurons in one direction only, from the dendrites to the axon. He showed that neurons communicate with each other at points of contact that he called *synapses*.

He intuited that information is transferred across synapses in one direction only, from the axon of an upstream neuron to a dendrite of a downstream neuron. He showed that specific functions, such as vision, are due to networks of neurons from different parts of the brain connected by long axons and synapses. Cajal's legacy to neuroscience cannot be overstated.

The flow of information in neurons consists of the *action potential*, a weak electrical current that diffuses in one direction only along the dendrites, cell body, and axon of the neuron. The electrical current is created by potassium and sodium ions moving across the neuron cell membrane through ion channels.

At synapses, the action potential triggers the release of small molecules called *neurotransmitters* in the small space that separates the two neurons. The neurotransmitters bind to their receptors on the other side of the synapse. This binding triggers an action potential in the dendrite of the downstream neuron.

The synapse can be compared to a dimmer switch that produces more or less light depending on the local circumstances. It causes an action potential in the downstream neuron only if it receives a strong signal from the upstream neuron. Synapses are not fixed structures. They can be pruned, and they can be remodeled—made stronger and more efficient at transmitting action potentials. Synaptic remodeling requires the synthesis of proteins.

Fluorescent reagents and in vivo microscopy allow imaging of the activity of single neurons in live laboratory mice. *Optogenetics* is a method that makes neurons fire action potentials when they receive short pulses of light delivered by a laser. Optogenetics can also be used to prevent neurons from firing. These recent techniques have revolutionized neuroscience, in particular the study of memory.

The immune responses in the brain are different from those in other organs. However, the brain is not an immune-privileged organ, one inaccessible to immune cells, as was believed for some time. Microglial cells constantly sample the brain and detect the presence of abnormal products, such as viruses and other intruders. Upon detection of the intruder, microglial cells trigger an immune response, including *inflammation*. Protracted inflammation is detrimental to neurons.

CHAPTER 4

A Prion Protein in Parkinson's Disease

Serendipity is a major player in science. After my move to Stanford, I was still going regularly to Paris to stay in touch with projects and colleagues. On a return trip to San Francisco, my suitcase was missing at the airport. "Don't worry, Mr. Melki," said the Air France clerk, "we will bring it to your home tomorrow."

"That's great, thank you very much," I said, "but I am not Mr. Melki."

"Ah! But your suitcase was checked in under Mr. Ronald Melki's name."

"That's not possible! Strangely, however, I do know a Mr. Ronald Melki very well; perhaps he was on the same flight."

That night I emailed Ronald, a colleague from Paris and an expert on prions. We used to meet when we were both sitting on a committee advising the French government on its handling of the "mad cow" epidemic. It turned out that he had just arrived at Stanford on sabbatical from his French laboratory. No, he did not have my suitcase; which Air France, notwithstanding, delivered to my doorstep the next day, as promised. Nobody ever understood how the Air France computer tagged my suitcase with Melki's name, but the airline mix-up changed my life as a scientist. Ronald and I made a date for lunch on campus and talked.

Ronald is an expert on prions and protein folding. His work at the time was dealing mainly with prions of yeasts, which we will consider later in this book, but he had recently shifted his interest to the prions responsible for human diseases. He explained to me that in 2008, Swedish and American neurologists had suggested that a protein called *alpha-synuclein* spread in the brains of Parkinson's disease patients like an infection.

The Neuron Lesions of Parkinson's Disease

Parkinson's disease is, unfortunately, common in people over sixty-five years of age. Along with Alzheimer's disease, it is one of the age-related neurodegenerative diseases that have become a major health concern in the aging population of the developed world. Parkinson's disease patients suffer mainly from rigidity of the whole body with impaired balance and tremors in their hands. However, several other debilitating symptoms, including dementia, may develop with time as the disease inexorably progresses.

In 1910, Friedrich Lewy, a German neurologist, was examining under the microscope sections of the brains of patients who had died of Parkinson's disease. He observed large abnormal structures inside neurons, which are now called *Lewy bodies*. He also noticed some thinner lesions inside axons, which

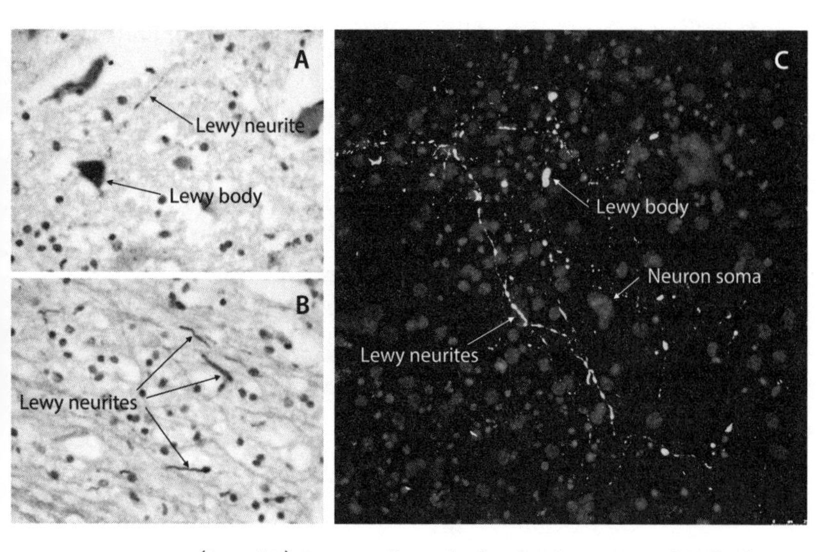

FIGURE 4.1. (A and B) Sections from the brain of a patient who died of Parkinson's disease. Lewy bodies and Lewy neurites were stained with an antibody specific for the alpha-synuclein prion protein. The antibody gives a specific color, which is lost in this black-and-white reproduction. (C) Human neurons in culture were incubated with seeds prepared from alpha-synuclein fibrils. After four days the cells were stained with the same antibody used in A and B. In this case the antibody gave a fluorescent color, which stands out against the dark background. An axon can be seen containing numerous Lewy neurites. A and B: From "Neuropathology of Sporadic Parkinson Disease before the Appearance of Parkinsonism: Preclinical Parkinson Disease" by Isidre Ferrer et al. (*Journal of Neural Transmission* 118 [2011]: 821–39); © 2011 Springer-Verlag

became known as *Lewy neurites*. Both are illustrated in figure 4.1. Lewy also reported that neurons with Lewy bodies looked unhealthy, and that some of them appeared to be dying. The presence of Lewy bodies and Lewy neurites became a hallmark of Parkinson's disease and is used by neuropathologists to make a diagnosis.

Lewy bodies are found in specific locations in the brain. They are prominent in a small region, deep in the brain, called the *substantia nigra*. The neurons in the substantia nigra make a chemical, a neurotransmitter called *dopamine*, which they send down their very long axons to a distant region of the brain called the *striatum*. Striatum neurons are involved in the control of body movements. When the neurons in the substantia nigra degenerate, the lack of dopamine in the striatum causes tremors and rigidity, the typical symptoms of Parkinson's disease.

In the 1990s, scientists examined the content of Lewy bodies and discovered that they were made of clumped proteins. The most abundant one was alpha-synuclein. This is a neuronal protein found primarily in synapses, the areas of contact between neurons where the impulse from the upstream neuron is transmitted to the downstream neuron. Alpha-synuclein plays a role in the complex mechanism by which neurotransmitters, such as dopamine, are released from the upstream neuron into the synaptic cleft. Using the electron microscope, scientists saw that in Lewy bodies alpha-synuclein formed long, hairlike filaments, which they called *fibrils*. Strikingly, these fibrils were reminiscent of those formed by the prion protein PrP in the brains of sheep with scrapie (chapter 1). With hindsight we now realize that this was a hint that alpha-synuclein in fibrils may be behaving like a prion.

The Spread of Lewy Bodies in the Brains of Parkinson's Patients

In 2003, Eva and Heiko Braak, two neuropathologists from the University of Frankfurt, published an influential article that summarized their work on the spread of Lewy bodies in the brains of patients with Parkinson's disease. The Braaks had assembled a large collection of autopsy material from Parkinson's patients who had died at different times after the onset of their disease. From a careful analysis of the Lewy bodies in this collection, they concluded that Lewy bodies appear to spread with time between brain areas that are connected by axons. They hypothesized that some unknown infectious agent was being transported in the axons, causing the formation of Lewy bodies and the demise of the neurons.[1]

At about the same time, neuropathologists in Sweden and in the United States made a surprising observation that pointed in the same direction. The standard treatment for Parkinson's disease is a drug given orally, called levodopa. The body turns levodopa into dopamine. The hope is that this will compensate for the deficit in dopamine in the striatum. It works well, but with time the effect of levodopa wears off, doses need to be increased, and bothersome side effects become a problem. Beginning in the 1990s, neurologists, particularly in Sweden and the United States, attempted to deliver dopamine directly to the striatum of the brain, where it is needed. To achieve this, they came up with the idea of grafting neurons that are able to make dopamine, obtained from aborted human fetuses, directly into the striatum of patients. Unfortunately, this was not very successful. The symptoms diminished in only a few patients. Also, it was technically

1. The research done by Eva and Heiko Braak had a major impact on our understanding of both Parkinson's and Alzheimer's disease. For a summary of their results on Parkinson's: H. Braak, U. Rüb, W. P. Gai, and K. Del Tredici, "Idiopathic Parkinson's Disease: Possible Routes by which Vulnerable Neuronal Types May Be Subject to Neuroinvasion by an Unknown Pathogen," *Journal of Neural Transmission* 110 (2003): 517–36.

difficult to obtain fetal neurons that make dopamine and graft them into the brain, and using human fetuses raised ethical questions. More than ten years later, when neuropathologists examined the brains of patients who had died after receiving a graft, they were surprised to find Lewy bodies not only in the substantia nigra and striatum, as expected, but also in the grafted neurons. It looked as if the new neurons had been "contaminated" by some product coming from the sick neurons. No one knew what this toxic product was. Among several hypotheses, they suggested a "prion-like" spread of alpha-synuclein aggregates.[2]

Testing the Prion Hypothesis for Alpha-Synuclein

Ronald and I discussed these findings and decided to join forces and test the hypothesis that alpha-synuclein behaves as a prion protein in Parkinson's disease. It would explain the observations made by Eva and Heiko Braak, as well as the presence of Lewy bodies in grafted neurons. The axonal transport of a prion protein, its transfer to new neurons, and the

2. The two articles reporting these results had a profound effect on our view of the spread of lesions in Parkinson's disease. They vindicated the hypothesis of Eva and Heiko Braak, based on their observations of the evolution of pathology over time in the course of the disease. The articles are: J. H. Kordower, Y. Chu, R. A. Hauser, T. B. Freeman, and C. W. Olanow, "Lewy Body–Like Pathology in Long-Term Embryonic Nigral Transplants in Parkinson's Disease," *Nature Medicine* 14 (2008): 504–6; J.-Y. Li, E. Englund, J. L. Holton, D. Soulet, P. Hagell, A. J. Lees, T. Lashley, et al., "Lewy Bodies in Grafted Neurons in Subjects with Parkinson's Disease Suggest Host-to-Graft Disease Propagation," *Nature Medicine* 14 (2008): 501–3.

mechanism of prion seeding and amplification described in chapter 2 could account for all of this. These were the exciting recent developments that Ronald and I discussed over lunch after Air France mislabeled my suitcase. If the hypothesis was correct, it might be possible to find drugs to interfere with the transport of alpha-synuclein prions or with their entry into new neurons. Such drugs could slow or halt the spread of lesions in the brains of Parkinson's disease patients. What kind of experiment should we perform to test the hypothesis?

At Stanford, we were trying to figure out how certain viruses spread from cell to cell in the brains of laboratory mice. Viruses can be very specific in the type of cell that they infect. They can also spread among brain cells by various routes, including axonal transport and cell-to-cell transfer through cell connections. For example, rabies virus spreads from neuron to neuron through the synapses that connect them. In the laboratory, we were using several different techniques to follow the spread of these viruses. We could look for viruses by staining brain sections with cell- and virus-specific reagents at different times after inoculation. However, this necessitated sacrificing the animals, and so did not allow us to follow what was happening in neurons carrying the virus in real time. Therefore, we were also using mouse neurons cultured in special devices that allowed us, using powerful microscopes, to observe the transport of the virus and its spread from neuron to neuron in real time in live neurons. During our first discussion it became clear to Ronald and to me that we should collaborate and use the tools that we had devised for mouse viruses to follow the traffic of the putative Parkinson prion. Indeed, we had complementary expertise. Ronald's people knew how to make large amounts of alpha-synuclein using genetically engineered bacteria. Remarkably, they had found ways of turning the protein, in vitro, into fibrils of the kind present in Lewy bodies. They could also attach a fluorescent chemical to the fibrils that made them visible under a special kind of microscope.

Ronald and I arranged to perform experiments together. First, we injected mice with fluorescent alpha-synuclein fibrils prepared by his laboratory and followed what happened in the brain with time. The experiments worked very well. Several weeks after the injection, we could detect large amounts of the fluorescent alpha-synuclein fibrils at the injection site, as expected, but also, and most interestingly, far away from the injection site, in areas connected to it by axons. These fibrils were inside neurons. This was the first indication that fibrils could indeed be transported long distances in axons and could be transferred to new neurons after transport. But we also wanted to find out what happened in these new neurons after the fibrils arrived. Did the fibrils act as seeds, causing the alpha-synuclein of the new neurons to form more fibrils? How could we determine that?

Scientists never work in isolation. They constantly need to be aware of what others are doing, not only in their own field but also in fields that are not obviously relevant to their own research. They do this by reading scientific journals, by going to meetings and seminars, and through informal discussions with colleagues. This is one reason why the environment in which one does one's research is so important. A high-profile research institute or university provides a considerable advantage, even while the internet is a remarkably efficient way of communication between distant laboratories.

In our case, we were helped by the work of other groups who discovered that alpha-synuclein undergoes a chemical modification in Lewy bodies, a modification that does not happen to normal neuron alpha-synuclein. The researchers had prepared a reagent, an antibody, that recognized only the chemically modified form of alpha-synuclein, and did not react with normal alpha-synuclein produced by genetic engineering. We obtained this antibody and were able to confirm that, in the test-tube, it did not react with the alpha-synuclein fibrils made in Ronald's laboratory, the ones we had injected into the brains of mice. We also performed several control experiments, which showed that the fibrils we injected were not modified in the brain—they were not stained by the antibody. Therefore, we were thrilled when we observed, one month after the injection, that the antibody stained Lewy-body-like and Lewy-neurite-like aggregates in neurons at the site of injection. The most likely explanation was that the injected fibrils were behaving like prion seeds, forming new fibrils, which were modified by the cell and became reactive to the antibody.

We were even more thrilled when we observed that the antibody reacted with neurons far from the site of injection, in areas connected to it by axons. This was very exciting. It suggested that the fibrils were being transported long distances in axons and could spread from these axons to new neurons and act as seeds in a prion-like mechanism. In brief, the simplest explanation for our results was that the fibrils that we injected behaved as prion proteins. They entered the local neurons and seeded the formation of more fibrils and of Lewy-body-like aggregates. They were transported over long distances in axons and entered new neurons, where again they acted as seeds, forming new fibrils and new aggregates.[3]

We were not alone doing these kinds of experiments. At about the same time, several laboratories, among them that of Virginia Lee at the University of Pennsylvania, were making similar observations.[4] Reproducibility is a very important aspect of science, especially in biology and medicine because of the large variations that may exist between groups of patients or between animal models. It is very encouraging when several laboratories obtain similar results using slightly different systems. In this case, all laboratories working in the field concluded that, at least in a mouse model, alpha-synuclein fibrils behaved like prions.

Up to that point, all our work had been done in mice. What about humans? Was it also true that misfolded alpha-synuclein behaved like a prion in the brains of Parkinson's disease patients? If true, this would change our view of the mechanism of the disease and open new possibilities to limit or stop the spread of the lesions. Obviously, the kinds of experiments that were possible with mice could not be done in humans. But,

3. Our work was published in several papers. Among them: E. C. Freundt, N, Maynard, E. K. Clancy, S. Roy, L. Bousset, Y. Sourigues, M. Covert, R. Melki, K. Kirkegaard, and M. Brahic, "Neuron-to-Neuron Transmission of α-Synuclein Fibrils through Axonal Transport," *Annals of Neurology* 72 (2012): 517–24; G. Bieri, M. Brahic, L. Bousset, J. Couthouis, N. J. Kramer, R. Ma, L. Nakayama, et al., "LRRK2 Modifies α-Syn Pathology and Spread in Mouse Models and Human Neurons," *Acta neuropathologica* 137 (2019): 961–80.

4. L. A. Volpicelli-Daley, K. C. Luk, T. P. Patel, S. A. Tanik, D. M. Riddle, A. Stieber, D. F. Meaney, J. Q. Trojanowski, and V. M.-Y. Lee, "Exogenous α-Synuclein Fibrils Induce Lewy Body Pathology Leading to Synaptic Dysfunction and Neuron Death," *Neuron* 72 (2011): 57–71.

fortunately, some remarkable advances in cell biology had taken place that allowed us to work on human neurons grown in culture. Where do such neurons come from? Strangely enough, they are obtained from a very small sample of human skin, a skin biopsy. To understand how this is possible, we need to introduce *stem cells*.

Stem Cells and Their Use in Medical Research

Stem cells appear in the press more and more frequently because of their promising applications in medicine. What are they, and how can they be used to study the prion proteins involved in neurodegenerative diseases?

Stem cells are central to development. For biologists, "development" refers to the process that turns a fertilized egg into an embryo and eventually into a complete individual, a baby in the case of humans. For a long time, embryology consisted in observing embryos under the microscope at different stages of their development and building a sort of a "movie" of cells multiplying, changing shape with time, moving along well-defined paths, and eventually forming limbs and other body parts as well as all the internal organs. With the arrival of molecular biology, many of the genes and molecules responsible for these changes, and the mechanisms that turn them on and off at the right time and in the right place, have been identified. However, there is still a lot to be discovered in development; the field is extremely active and important for medicine. It is relevant not only to the genetic malformations that can occur during development but also to the causes of many diseases that occur later in life, including cancer.

In the embryo, stem cells are a group of cells that keep dividing, producing more stem cells, and that do not look like the cells of any specific tissue, say neurons or liver cells. For this reason, they are said to be *undifferentiated*. From this group of dividing stem cells, some emerge, at certain times and in certain places, change shape, eventually stop dividing and turn into the various specialized cells that make our organs, including the neurons in our brain.

Embryonic stem cells are *pluripotent*, meaning that they have the potential to differentiate into multiple different cell types found in the body. Our understanding of the subtleties involved in turning a pluripotent stem cell into a specialized cell such as a neuron is far from complete. However, it is now possible to obtain pluripotent stem cells from embryos, in practice mainly mouse embryos, to grow them in culture, and to turn them into some of the specific cell types that make a body, including neurons. This is a major advance since it allows one to perform experiments in simple cell culture, outside of the organ and the animal.

In the case of humans, working with stem cells from embryos produced by in vitro fertilization raises ethical issues and is subject to regulations that strongly limit their use. Some years ago, however, a remarkable advance took place that made these restrictions obsolete. In 2006 Shinya Yamanaka at Kyoto University reported that it was possible to obtain pluripotent stem cells from adult, differentiated cells, including from cells obtained from a skin biopsy.[5] In other words, scientists realized

5. Yamanaka's publication had a profound influence in biology: K. Takahashi and S. Yamanaka, "Induction of Pluripotent Stem Cells from Mouse Embryonic and Adult Fibroblast Cultures by Defined Factors," *Cell* 126 (2006): 663–76.

that it was possible to "de-differentiate" specialized cells into stem cells; to turn back the clock of development. To do this, it is necessary to turn on genes that are specific for embryonic pluripotent stem cells. The cells obtained in this way are called *induced pluripotent stem cells,* or iPSCs for short. These iPSCs can be grown in culture in the laboratory. Like stem cells obtained from embryos, they can be differentiated into various types of adult cells, again including neurons. Being able to obtain iPSCs from a simple skin biopsy and to differentiate them at will opened entirely new possibilities in cell biology. In medicine it opened new avenues to study the mechanisms of many diseases. For example, it is now possible to compare neurons from a patient whose neurodegenerative disease is caused by a mutation in a particular gene with neurons from an individual without the mutation, and even with neurons from the same patient in which the mutation has been corrected by genetic engineering. Those comparisons are important because they can lead to treatments not only for patients with the mutation but also for patients whose disease is not directly caused by a mutation.

Using Human Induced Stem Cells to Study Parkinson's Disease

Several laboratories, including Ronald's and mine, used iPSCs to prepare human neurons. In these cultures neurons extend very long axons, establish contacts between one another, and send electrical signals down their axons. This is obviously an extremely simplified form of brain tissue, but it nevertheless allows observation of what happens when live, active human neurons are exposed to various agents, including alpha-synuclein fibrils.

We added fluorescent alpha-synuclein fibrils to such cultures and followed their behavior over time under the microscope. The first thing we observed was that the alpha-synuclein fibrils entered the neurons. This was interesting because cells are very particular about what they take in from their environment. As a rule, a protein needs to be recognized by a specific receptor on the surface of the cell to be taken up. We still do not know if there is a receptor for alpha-synuclein fibrils or the mechanism of entry into neurons. These are important questions and are being studied by several laboratories. Knowledge of the receptor and mechanism of entry will help in the design of drugs to interfere with the spread of lesions in the brains of Parkinson's disease patients.

Once the fibrils were inside the neurons, we observed that they were transported over considerable distances in the axons. This was also interesting, since axons do not transport just anything that happens to be there; there is specificity there too. Here again, examining the mechanism of transport in detail could lead to drugs that limit the spread of the disease.

Of particular interest was the presence of aggregates stained with the prion-specific antibody in the axons of neurons that had ingested the fibrils. It appeared that the normal, *endogenous* alpha-synuclein of axons had been turned into a prion form, recreating in culture the typical Lewy neurites seen in the brains of Parkinson's disease patients (see figure 4.1C). Pathologists using the same antibody to stain brain sections from Parkinson's disease patients have now shown that most of the aggregated alpha-synuclein is in Lewy neurites in axons, not in the large Lewy bodies. Since this is precisely what we observed in our human iPSC-derived neurons, it makes such cultures a good model to test drugs that will interfere with the propagation of the lesions of Parkinson's disease.

How Does Alpha-Synuclein Turn into a Prion in the Brains of Parkinson's Patients?

You are probably wondering how Parkinson's disease starts in the brain of a patient. We all have neurons that contain a lot of alpha-synuclein, yet only some of us are unfortunate enough to come down with Parkinson's disease. How and why does alpha-synuclein, an important neuron protein, turn into a prion that spreads through the brain and kills neurons? To understand this we need to go back to the phenomenon of protein folding, which we considered in some detail in chapter 2.

Alpha-synuclein is an *intrinsically disordered protein*. You may remember from chapter 2 that intrinsically disordered proteins do not fold rapidly into a final, stable, specific shape after they are made. They persist inside the cell as unfolded or only partially folded chains of amino acids. In fact, they constantly fold and unfold in many different shapes, which are all unstable. All of this happens very quickly, in only a few milliseconds, a few thousandths of a second. Obviously, a protein that behaves like this cannot be much use to the cell. Intrinsically disordered proteins eventually do fold and acquire a stable shape, but only after they find and bind to a specific component, called a *partner*. For many of them the partner is another protein with which they form a stable duo. For alpha-synuclein the partner is the surface of a membrane. Binding to a membrane gives alpha-synuclein its normal, final shape.

Like everything, this process can sometimes go wrong; proteins can *misfold*. They fold in more-or-less stable shapes that are different from their normal, functional shape. Because this is dangerous for the cell, cells are equipped with a process that we discussed in chapter 2, called *protein quality control*. Misfolded proteins are detected and either refolded or degraded. However, with age quality control becomes less efficient, and some misfolded proteins are not eliminated. Decline in quality control is an important factor in many age-related diseases, including neurodegenerative diseases such as Parkinson's and Alzheimer's.

As we explained in more detail in chapter 2, *self-templating* is what makes a protein a prion. Self-templating means that the misfolded protein will cause other copies of the same protein to misfold into the same shape. If alpha-synuclein misfolds into a prion protein, it will give the same prion shape to other copies of alpha-synuclein present in the same neuron. How this works is still the subject of ongoing research, and not everybody agrees on a detailed mechanism. Figure 2.4 in chapter 2 illustrates a likely scenario. Self-templating creates a pile of identical misfolded protein molecules that gets longer and longer with time, eventually making a fibril, which can be seen with the electron microscope. This fibril can get broken into pieces, which function as seeds, initiating the formation of new fibrils. As time goes on, many fibrils may clump together, as well as with other proteins, forming Lewy neurites and Lewy bodies. If some alpha-synuclein fibrils, or pieces of fibrils, are released from a neuron and enter another neuron, seeding and fibril formation will repeat in this new neuron, which will eventually contain new Lewy neurites and Lewy bodies. With time the process will involve more and more neurons. Lewy bodies will spread through the brain like an infection.

You may think that this represents a terrible flaw in our brain, and you may wonder why we don't all have Parkinson's disease. This is in part a question of chance. The unstable, misfolded alpha-synuclein molecules have a very short life, on the order of a few milliseconds. Therefore, the chance that two misfolded alpha-synuclein molecules that happen to have the same prion shape meet inside the same neuron forming a stable pair and triggering self-templating is extremely small. In many instances, these misfolded forms are detected by the quality control system of the neuron and are refolded or destroyed. It takes both bad luck and poor protein quality control to initiate the formation of alpha-synuclein fibrils in a neuron. But besides bad luck there are other factors that contribute to Parkinson's disease.

The risk of developing Parkinson's disease can be affected by genetic factors. Mutations in the gene for alpha-synuclein can cause the disease in all family members who carry the mutation. These cases prove that alpha-synuclein *can* be a cause of the disease. They strongly point to Lewy bodies and Lewy neurites, which are composed of misfolded alpha-synuclein, as causing the death of neurons. However, these familial cases are very rare. But there are more common genetic factors, and close to a hundred of them have been detected. These mutations have only a modest effect—they only slightly increase the risk of developing the disease.

Besides bad luck, age, and genetic factors, toxic products in the environment, possibly some herbicides and pesticides, appear to be involved. Some infections that disrupt the immune response to abnormal proteins may also play a role. It is very likely that there are environmental factors for Parkinson's disease that have yet to be identified. Because of such complexity, we speak of Parkinson's disease, and many other neurodegenerative diseases, as *multifactorial.* They are complex, difficult to understand in all their details. However, working out these details is essential in order to develop drugs for treatment and hopefully prevention.

There Are Strains of Alpha-Synuclein Prions

There is yet another fascinating twist in the story of alpha-synuclein prions. Besides Parkinson's disease, alpha-synuclein clumps are observed in two other neurological diseases: dementia with Lewy bodies and multiple system atrophy. Whereas Parkinson's disease is relatively common, these other two are rare. As its name indicates, dementia with Lewy bodies is a form of Parkinson's disease in which dementia is the main symptom. Lewy bodies are found in many areas of the brains of these patients, including the cortex, its outermost layer.

The symptoms of multiple system atrophy are very different from those of Parkinson's disease and dementia with Lewy bodies. This difference reflects the fact that the alpha-synuclein clumps are not in neurons but in oligodendrocytes, the glial cells that make myelin. Parkinson's disease, dementia with Lewy bodies, and multiple system atrophy are now collectively called *synucleinopathies.*

The existence of three different diseases associated with alpha-synuclein aggregates creates a new problem. If alpha-synuclein aggregates are the cause of these diseases, how can three different diseases have the same cause? Maybe alpha-synuclein clumps are not the cause of the diseases after all, but simply a common response of the cells to injuries inflicted by three different, and still unknown, causes. Maybe alpha-synuclein aggregates cause different diseases depending on the genetic background of the patient. Solving this problem led to a remarkable conclusion.

Several laboratories extracted the alpha-synuclein fibrils from the brains of patients with the different diseases. They observed that, under the electron microscope, the fibrils had different shapes. Remember that the fibril is a regular stack of misfolded protein molecules, each molecule with the same shape. In chapter 2, we compared this stack to a pile of dishes. Looking carefully at the shape of each protein in the stack showed that the shape was different in the three different diseases. Like piles of soup dishes versus piles of desert dishes. The different shapes were due to the protein having misfolded in different ways. In other words, it looked as though alpha-synuclein could misfold in three different ways, giving rise to three different kinds of prions, each one responsible for a different disease. This hypothesis was tested directly by injecting the different fibrils into the brains of mice that had been genetically modified to express the human alpha-synuclein gene. The mice developed different diseases.[6]

6. J. I. Ayers, J. Lee, O. Monteiro, A. L. Woerman, A. A. Lazar, C. Condello, N. A. Paras, and S. B. Prusiner, "Different α-Synuclein Prion Strains Cause Dementia with Lewy Bodies and Multiple System Atrophy," *Proceedings of the National Academy of Sciences of the USA* 119 (2022): e2113489119.

Strains of viruses that give slightly different diseases are well known to geneticists. They are due to mutations in a gene of the microbe, and these strains are usually called *variants*. The Covid epidemic has unfortunately made us familiar with the concept of mutations as responsible for the appearance of new variants. But in the case of Parkinson's disease, dementia with Lewy bodies, and multiple system atrophy, there are no mutations in the gene for alpha-synuclein that are responsible for the different diseases. The differences are due to the different shapes taken by a protein. From these studies we learned that mutations are not the only explanation for the appearance of variants. There is a "new" genetics, besides classical genetics based on DNA, RNA, and mutations. The central dogma of biology is being overturned by the study of prion diseases. Genetics is more complicated than we had thought. Prion proteins are part of what is now referred to as *epigenetics*.

The discovery that a protein with prion property has a central role in Parkinson's disease opened a new chapter in neurology. In fact, as predicted by Stanley Prusiner many years ago, prion proteins are involved in several neurodegenerative diseases, including the much-feared Alzheimer's disease, which we will examine in the next chapter.

New Treatments for Parkinson's Disease

Levodopa, a drug that the body turns into dopamine, has been used for decades to treat Parkinson's disease, to correct the lack of dopamine caused by the death of neurons in the substantia nigra. Levodopa is very useful and successful, but with time its side effects become too great and, unfortunately, it does not halt the relentless progression of the disease or the spread of lesions to more and more distant regions of the brain. It is only a replacement therapy.

Cell replacement therapy, which in the case of Parkinson's disease consists in grafting dopamine-producing neurons into the brains of patients, has not been completely abandoned. In fact, it is still an active field of research. Instead of using human embryos as a source of these neurons, neurologists take advantage of recent developments in producing human-iPSCs, which can be differentiated into dopamine-producing neurons. Clinical trials are in progress with this new approach.

As already mentioned, the discovery that alpha-synuclein is responsible for the formation and the spread of Lewy bodies and Lewy neurites opened new avenues for treating the disease. Obviously, preventing self-templating of alpha-synuclein, blocking the spread of alpha-synuclein prions from neuron to neuron, and removing the toxic alpha-synuclein aggregates from neurons should all be beneficial to patients.

Alpha-synuclein targeted therapies are now at the front line in the search for Parkinson's disease drugs. Scientists screen collections of thousands of already-existing molecules for any putative activity on self-templating, spread, or removal of alpha-synuclein aggregates. The screening is done on neurons in culture, including human neurons obtained from iPSCs, as discussed above. It is also done on animal models of the disease, mainly transgenic mice that spontaneously develop alpha-synuclein aggregates with time, and mice injected with alpha-synuclein prions.

Drugs aimed at the prion form of alpha-synuclein may also be obtained by drug design. The remarkable amount of knowledge gained on the atomic structure of alpha-synuclein in its prion form, and on the mechanism of self-templating, enables chemists to design molecules that are predicted to interfere with the formation of these toxic components. Such molecules are synthesized by chemists and tested for their activity. This research, which is very active, has not yet produced any molecules ready to enter clinical trials, but it carries great hopes.

Clinical trials are ongoing with antibodies against alpha-synuclein. These antibodies are proteins that very specifically bind to alpha-synuclein, some of them only to the aggregated, fibrillar form of the protein. They might prevent the entry of alpha-synuclein prion seeds into neurons. The hope is that they will intercept alpha-synuclein prions as they diffuse between neurons, "infecting" healthy neurons and spreading the disease in the brain.

Another approach to treatment consists in boosting protein quality control in neurons to eliminate aggregated alpha-synuclein. Protein quality control is a series of complex mechanisms that refolds or degrades misfolded proteins. This normal activity of cells, including of neurons, tends to decrease with age. Drugs that boost protein quality control already exist and are being studied in the broader context of age-related diseases. One of them is called rapamycin. Ongoing research aims at finding more drugs that are more specific to diseases and less toxic.

To Recap

Serendipity is important in research. It led me to discuss the prion hypothesis for Parkinson's disease with an expert in yeast prions and to set up a collaboration between the two of us.

The diagnostic lesions in Parkinson's disease are *Lewy bodies* and *Lewy neurites*. These abnormal structures are associated with neuron death. Lewy neurites and Lewy bodies are made of aggregated proteins, in particular *alpha-synuclein*. They are toxic for neurons.

Neurons in an area called the *substantia nigra* produce the neurotransmitter *dopamine*. In Parkinson's disease these neurons contain Lewy bodies and Lewy neurites. They malfunction, then die. The lack of dopamine resulting from their death is responsible for patients' tremors.

Fetal neurons producing dopamine were grafted into the brains of Parkinson's disease patients to treat the disease. At autopsy, more than ten years later, Lewy bodies were found in the grafted neurons, suggesting that misfolded alpha-synuclein could spread from cell to cell like an infectious agent.

Eva and Heiko Braak showed that Lewy bodies spread over time in the brains of Parkinson's disease patients, between areas connected by axons. They suggested that this observation could be explained if alpha-synuclein, the main component of Lewy bodies, behaved as a prion.

We tested the prion hypothesis by injecting alpha-synuclein fibrils into the brains of mice and by using iPSC-derived human neurons in culture. We observed that the fibrils did behave as prion seeds. They triggered the misfolding of the neurons' normal alpha-synuclein into more prions. They were transported in axons and transferred to more neurons. Similar findings were made by several laboratories. The prion nature of misfolded alpha-synuclein is now generally accepted.

The misfolding of alpha-synuclein in the brain and the development of Parkinson's disease is a rare, random event that may be favored by a decline of *protein quality control* with age and possibly by some environmental factors such as exposure to toxic chemicals. Parkinson's disease is a *multifactorial* disease. These diseases are difficult to study, but it is important to decipher their mechanisms in order to design drugs that halt their development and treat them and to implement prevention.

Two other diseases besides Parkinson's, are caused by alpha-synuclein misfolding into a prion protein. They are *dementia with Lewy bodies* and *multiple system atrophy*. In each case the shape taken by the alpha-synuclein prion protein is different. Alpha-synuclein can fold into at least three different prion proteins, which correspond to "strains" of alpha-synuclein prions. These strains do not arise through mutations in the gene coding for the protein but through different shapes taken by the protein. The existence of strains due to differences in folding is a new concept in genetics. Prion proteins are part of what is now called *epigenetics*.

Alpha-synuclein in its prion form is now a main drug target for the pharmaceutical industry. Small molecules as well as antibodies are being tested for their activity on self-templating, spread, and toxicity of the aggregated form of alpha-synuclein.

CHAPTER 5

Two Prion Proteins Are Responsible for Alzheimer's Disease

Alzheimer's disease is a progressive dementia, a slow, relentless decline in cognitive function. The disease is now so common in the aging human population that almost all of us know a sufferer or have one among our family or friends. Since age is its main risk factor, as people live longer the number of patients increases, causing a considerable emotional and financial burden to their families and to society in general.

One of the most debilitating symptoms of Alzheimer's disease is the progressive loss of memory, especially short-term, working memory. The memory of what just happened, of where we are, the memory that is needed to plan immediate actions; the memory that makes everyday life in society possible. When short-term memory becomes severely impaired, which invariably happens after some years of disease, patients are unable to perform even ordinary, everyday tasks such as getting dressed or making a cup of tea. Eventually they cannot look after themselves and become entirely dependent on external help.

The recall of long-term memory, the memory of long-gone events, of our childhood, unfortunately will also be affected by Alzheimer's disease, although less rapidly than the acquisition of short-term memory. However, losing long-term memory is a real tragedy. We are who we are because of our genetic inheritance, but also in great part because of our memories, as since childhood our brain has been storing, mostly in an unconscious form, an enormous amount of information about past events. A duality often referred to as nature versus nurture. Losing nurture, losing long-term memory, robs us of a large part of our personality.

In 1906 Alois Alzheimer, a German neurologist and pathologist, gave a talk at the 37th Congress of Psychiatrists of Southern Germany, during which he described the lesions he had observed in the brain of one of his patients. The patient had died after developing progressive dementia over several years. Alzheimer reported seeing a considerable loss of brain substance in the gray matter of the cortex, the layer of neurons at the surface of the brain. Under the microscope, this loss was explained by a diminution of the number of neurons. Alzheimer also described abnormal aggregates of material present in between neurons. These aggregates had been observed before in elderly individuals and were known as "senile plaques." They have now been renamed *amyloid plaques*. What had not been noticed before, which Alzheimer described in detail, were abnormal structures present inside neurons, which he called *neurofibrillary tangles*. Plaques and neurofibrillary

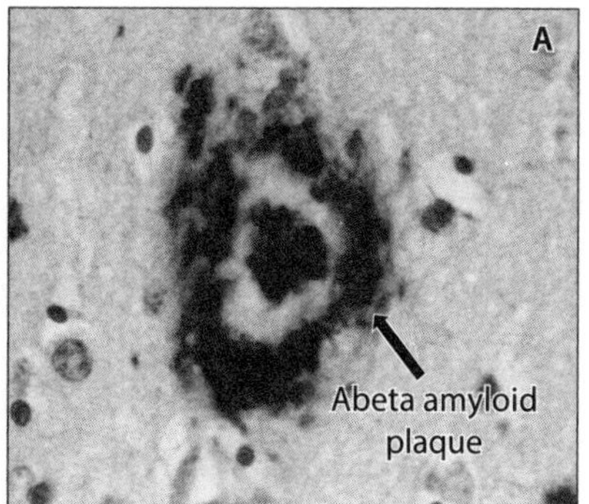

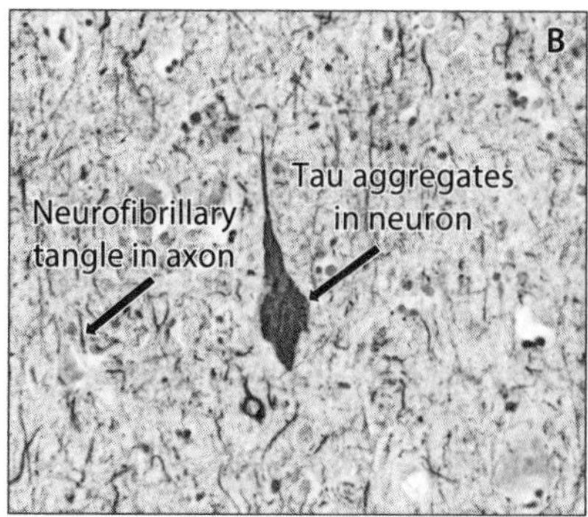

FIGURE 5.1. Sections from the brain of a patient who died of Alzheimer's disease. (A) High magnification of an amyloid plaque stained with an antibody against the A-beta prion protein. (B) The section was stained with an antibody against the tau prion protein. Numerous neurofibrillary tangles show up, as well as one neuron body containing tau prion protein aggregates. From *Neuropathology Simplified: A Guide for Clinicians and Neuroscientists* by David A. Hilton and Aditya G. Shivane; © 2015 Springer International Publishing Switzerland

tangles are still the hallmarks of the disease for pathologists (figure 5.1).[1]

The study of the plaques in Alzheimer's disease took a decisive turn in the 1920s when Paul Divry at the University of Liège in Belgium showed that the plaques contained what pathologists used to call an "amyloid substance." The chemical

1. The history of Alzheimer's disease research is remarkably well presented by Thomas Beach in "A History of Senile Plaques: From Alzheimer to Amyloid Imaging" (*Journal of Neuropathology and Experimental Neurology* 81 [2022]: 387–413) and by L. C. Walker in "Prion-Like Mechanisms in Alzheimer's Disease," chapter 16 of *Human Prion Diseases*, edited by M. Pocchiari and J. Manson (Handbook of Clinical Neurology vol. 153; Amsterdam: Elsevier, 2018).

nature of amyloids remained the subject of an animated controversy until it was finally shown that amyloids were made of proteins particularly rich in beta strands (you may remember from chapter 2 that the backbone of proteins consists in alpha helices and beta strands). Finally in the early 1980s two research groups, that of David Allsop at the University of Nottingham and that of George Glenner at the University of California, San Diego, identified the protein that made the amyloid substance of the Alzheimer's plaques, which became known as *amyloid-beta, Abeta* or Aβ. It is a small protein, forty to forty-two amino acids long. In fact, it is a fragment, called a *peptide,* of a larger protein named *amyloid precursor protein* (APP). APP is exposed at the surface of neurons, on the outside of the cell. The Abeta peptide is produced by enzymes that cut the chain of amino acids of the APP protein. Several Abeta peptides clump together and form aggregates outside the neurons. Eventually, the aggregates, together with some other proteins, form the amyloid plaques that are visible under the microscope.

The tangles, on the other hand, are mostly made of aggregates of a protein called tau. Tau is a well-studied protein involved in the transport of material inside cells. It interacts with microtubules, which are tracks throughout cells along which cellular motors carry cargo, such as proteins and RNAs. In neurons such motors and their cargo move constantly up and down the dendrites and the axon.

Abeta and Tau Are Prion Proteins

Around the same time that they were studying Parkinson's disease, Eva and Heiko Braak were also studying the brains of Alzheimer's disease patients. Their approach was similar in both cases. They assembled a large collection of specimens and did a careful study of the pathology of the disease in patients who died at various times after its onset. As research on Abeta and tau progressed, they were able to stain the specimens with antibodies specific for the aggregated forms of Abeta and tau. Their results painted a picture that was more complicated than that for Parkinson's disease. As other pathologists had noted, there was no Alzheimer's brain without Abeta aggregates. However, Abeta amyloid plaques were also present in the brains of people without Alzheimer's, in controls. We now know that such plaques are a common finding in individuals over seventy years of age. There was no clear correlation between the amount of aggregated Abeta and the clinical progression of the disease. In contrast, the amount of neurofibrillary tangles and tau aggregates correlated well with the progression of the disease. These observations, confirmed by others, formed the basis for a largely accepted scheme, according to which Abeta aggregates are necessary for the disease, they may somehow trigger the aggregation of tau, and the tau aggregates are directly responsible for dementia. This may be an oversimplification, and we will come back to the interactions between Abeta and tau later in this chapter.

Critically, the stages of the disease described by Braak paralleled the spread of aggregated tau in the brain over time.[2] Aggregated tau first appeared in an area called the *entorhinal cortex*. This is an area involved in the acquisition of memory

2. These articles by Heiko and Eva Braak became classics of the neurology literature: H. Braak and E. Braak, "Neuropathological Stageing of Alzheimer-Related Changes," *Acta neuropathologica* 82 (1991): 239–51; H. Braak and E. Braak, "Staging of Alzheimer's Disease-Related Neurofibrillary Changes," *Neurobiology of Aging* 16 (1995): 271–78.

and navigation through space and time, remembering where one was at a given time. It is a relay between the *hippocampus,* one of the first areas to process the acquisition of memory, and the *neocortex,* where long-term memory is eventually stored. From the entorhinal cortex, the Braaks noticed that tau aggregates spread to many other areas of the brain, all of them connected by axons. This pattern led them to define several stages of the disease, according to the extent of the lesions. The pattern is very different from the stages of Parkinson's disease.

Heiko and Eva Braak interpreted their results to mean that Abeta was somehow involved in the aggregation of tau in the entorhinal cortex, and that from there a toxic product responsible for tau aggregation spread by axonal transport. In 2011, Heiko Braak and his collaborator Kelly del Tredici hypothesized that the toxic product was the aggregated tau protein itself, and that it self-aggregated by a prion-like mechanism.

That tau is a prion protein was convincingly demonstrated by experiments in mice and in neuron cell culture, experiments that were very similar to those performed with alpha-synuclein in Parkinson's disease. Several laboratories, including those of Michel Goedert at the Medical Research Council's Laboratory of Molecular Biology in Cambridge in the United Kingdom and of Markus Tolnay at the University of Basel, injected tau fibrils extracted from the brains of patients into the brains of mice that were transgenic for the human tau gene. They showed that the fibrils acted as seeds for the aggregation of the human tau protein. Furthermore, they observed the spread of the tau-containing lesions from the site of injection to distant regions connected by axons.

For Abeta, convincing evidence that it was a prion protein was obtained in several laboratories, including those of Mathias Jucker (University of Tubingen), Claudio Sotto (University of Texas, Houston), and Lawrence Walker (Emory University). They used mice that were transgenic for the human *APP* gene, the gene that codes for the protein that is cleaved to produce the Abeta peptide. These mice do not develop a disease spontaneously and do not have plaques in their brains, even in old age. However, when they were injected with proteins extracted from the brains of Alzheimer's disease patients, amyloid plaques appeared at a distance from the site of injection. If Abeta was eliminated from the brain extract using antibodies, the plaques did not appear. Furthermore, plaques developed in mice injected with Abeta aggregates prepared in vitro from pure Abeta obtained by genetic engineering. And finally, the plaques could be transmitted from mouse to mouse by intracerebral injection of brain extracts.

Which Causes Disease: Abeta, Tau, or Both?

This is obviously an important question, and very relevant to treatment design. You need to know who the enemy is before you can make a plan of action. The good correlation between the spread of tau aggregates and the progression of the disease suggests a role for tau in neuron death and dementia. But the lack of such correlation with disease in the case of Abeta does not rule out its also having a role. Indeed, the Braak model suggests that Abeta aggregates might trigger tau aggregation. How did scientists proceed to untangle such a complicated situation?

First, the original 1991 work of the Braaks in elucidating the stages of disease progression, which had been done using autopsy material, was updated using new methods of in vivo imaging. These methods allow following the spread of Abeta and tau aggregates in the brains of patients during the course of their disease. The main method is *positron emission tomography* (PET). In PET, a short-lived radioactive atom is attached to a molecule, called the *marker*, that binds specifically to the target, in this case aggregated tau or aggregated Abeta. The marker can be any nontoxic compound, such as an antibody, that specifically recognizes its target. The marker when bound to its target is detected by a machine that scans the entire brain and constructs a three-dimensional image of the position of the target within it. Several markers have been designed, both for aggregated Abeta and aggregated tau.

The results of PET scans confirmed and complemented the results obtained by the Braaks. They showed that Abeta aggregates appear first. They confirmed that the spread and the amount of tau aggregates correlate rather well with the clinical status of the patient, with the extent of dementia, whereas the amount of Abeta does not. They also confirmed that after a certain age, many people who will not develop Alzheimer's disease have Abeta, and even tau, aggregates in their brains. PET scans are now of great clinical importance, since they allow the monitoring of the progression of the disease. They are also very helpful in testing the efficacy of candidate drugs during clinical trials.

Second, genetics has provided important clues on the respective roles of Abeta and tau. How was this achieved? There are familial cases of Alzheimer's disease, cases that occur in several generations of the same family. They are rare and the symptoms usually appear early in life, but the lesions and their progression are the same as in the common, sporadic form of the disease. Therefore, what one learns from familial cases is most likely relevant to the common form of the disease. So how do scientists use such familial cases to find the cause of a disease?

The first step is performed by neurologists. They examine medical records, and, when possible, also patients, to confirm the diagnosis of Alzheimer's disease. Of course, some patients have died, some are alive but unavailable, the medical records are not always accurate. All this limits the number of families that can provide useful genetic information. Since familial cases of Alzheimer's disease are rare, these genetic studies require collaborations between several medical centers, often in different countries. The next step is to collect DNA from all available members of the families. Then the work moves to the laboratory and is greatly helped by powerful, and relatively inexpensive, techniques to sequence DNA. Geneticists look for a mutation that is present in the DNA of all the Alzheimer's disease patients from a given family. If they find one, they determine if the mutation has been transmitted among family members according to Mendel's laws of heredity. When a mutation satisfies these criteria, one can be virtually certain that the product of the gene, the encoded protein, is causing the disease. This is extremely important information.

Three genes have been identified in this way in different families with Alzheimer's disease. Their abbreviated names are *APP*, *PSEN1*, and *PSEN2*. *APP* codes for the APP protein, the protein that gives the Abeta peptide. *PSEN1* and *PSEN2* code for the enzymes that cleave APP and release the Abeta peptide. Therefore, three different genes that cause familial Alzheimer's disease when they are mutated are all involved in the formation of Abeta aggregates. This is a very strong argument in favor of a central role of these Abeta aggregates and the amyloid plaques that they form in the mechanism of Alzheimer's disease. Even though the amount and spread of Abeta aggregates does not correlate with dementia, it justifies developing drugs to limit the formation and the spread of Abeta peptide or to remove it from the existing amyloid plaques.

Another genetic strategy, the *genome-wide association study* (GWAS), can also help to find the cause of complex diseases such as Alzheimer's. The technique relies on DNA sequencing and sophisticated statistical methods. It examines a very large number of patients and unaffected controls and looks for statistically significant differences between the DNA sequences of patients and controls. The question is whether some sequences are found more often in the patients than in the controls. It sounds simple, but identifying DNA sequences that have a high probability of being associated with the disease is not trivial at all. Such sequences identify what geneticists call *risk factors*, genes that are not strictly speaking causal but are associated with an increased risk of developing the disease. Such genes may have a small effect, but they are part of the metabolic processes involved in causing the disease. As such, the GWAS strategy also provides very useful information.

For Alzheimer's disease, such studies have detected mutations associated with high risk in the gene coding for the tau protein. This confirmed a role for tau, and so, following the same reasoning as for Abeta, tau aggregates are also potential drug targets.

The GWAS approach has provided another, unexpected, important piece of information. It uncovered that inflammation plays a part in Alzheimer's disease. Inflammation is part of the immune response to invading microbes or cancer cells. In the brain, invaders are detected by the microglial cells (see chapter 3). These cells are present everywhere in the brain. They are very active; they constantly sample their environment looking for the presence of abnormal molecules and structures. They detect amyloid plaques, engulf fragments of the plaques, and degrade them. This is a beneficial cleanup, but it can also be counterproductive. Besides cleaning up, microglial cells trigger an immune response against what they have ingested. They send chemical signals to attract immune cells, *lymphocytes,* which circulate in the blood and can enter the brain. They also secrete toxic molecules aimed at fighting microbes. Inflammation is beneficial in acute infections, infections that do not last; it participates in the elimination of microbes. But because Abeta and tau aggregates are not foreign microbes that can be eliminated from the body but proteins of the patient turned into prions, they are constantly replenished and are never eliminated. Abeta and tau are always present in the brain and available to form new prion aggregates. So the inflammation persists, and the toxic molecules secreted by inflammatory cells become toxic for the neurons in the vicinity. Acting on inflammation in the brain with drugs is the object of much research. The goal is to help microglial cells to remove plaques while limiting the damage due to inflammation.

The question we asked at the beginning of this section was which is the cause of Alzheimer's disease, Abeta or tau? The answer is both. However, they may not act independently of each other. In fact, they may interact in several ways. For example, as suggested by pathology and PET studies, Abeta aggregates may trigger the transformation, the misfolding, of tau into prion proteins. If this is confirmed, we need to understand the mechanism of such *cross-seeding*, the induction of tau self-templating by Abeta seeds, which have a totally different sequence of amino acids. Cross-seeding is a hotly debated topic in the field of protein folding. Some experts think that it is not possible, that it does not exist. The question is important not just for Alzheimer's disease. If cross-seeding is a reality it has implications for several other neurodegenerative diseases, which will not be dealt with in this book, in which two or more prion proteins appear to interact to cause neuron death.

Are Abeta and Tau Prions Toxic for Neurons?

A large amount of data, obtained mainly with neurons in culture, indicates that Abeta aggregates are toxic—not necessarily the large fibrils but the small ones, those with a limited number of Abeta molecules, which are found inside neurons. Why are they toxic? Neurons need a large amount of energy to perform their complicated functions. Like all cells, they obtain this energy from the oxygen carried by red cells in the blood. In cells this oxygen is transformed into chemical energy in specialized organelles called *mitochondria*. Experiments done in different laboratories show that Abeta aggregates poison mitochondria, thereby restricting the amount of energy available to the neuron. The amyloid plaques and their beta aggregates are located outside neurons. How small aggregates enter neurons is not clear at present.

Tau is a protein involved in the transport of molecules and organelles, including mitochondria, along dendrites and axons.

Tau aggregates, which form the neurofibrillary tangles described by Alzheimer inside neurons, appear to interfere with the transport of these essential components. Obviously, this will have deleterious consequences for the functioning of the neuron, in particular at the level of synapses, which receive most of their needs by axonal transport from the cell body. Both for Abeta and for tau, more research is needed to understand why they are toxic, and to try to fight their toxicity.

Abeta and Tau Aggregates Are Prion Proteins: What Are the Implications for Treatment?

What can be done to deal with these culprits? The first obvious idea is to remove them. This is precisely what several clinical trials are trying to achieve with antibodies. For Abeta, after several disappointments, the trials have now given positive results, and at least one antibody has been approved for clinical use in the United States. This is certainly not the end of the story. Better and better antibodies will certainly be found. Antibodies against aggregated tau are also being tested, although the trials are less advanced than for Abeta. Removing tau aggregates, which unlike aggregated Abeta are inside neurons, may be difficult. Despite that, at least two clinical trials are ongoing at the time of writing.

What makes Alzheimer's disease such a dreadful condition is that the Abeta and tau aggregates spread with time to more and more regions of the brain. It is the accumulation of neuron dysfunction and death in more and more areas of the brain that causes dementia. Halting the spread as early as possible, ideally before confirmed dementia, would be very beneficial.

Since the spread and accumulation of neuron dysfunction are due to Abeta and tau being prion proteins, treatments could address at least two distinct steps: self-templating and transport. Self-templating produces more and more seeds, and transport, spreads the seeds across the brain.

Blocking self-templating with drugs has been attempted for PrP, the scrapie agent. But the "mad cow" epidemic is over, Creutzfeldt-Jakob disease is rare, and the pharmaceutical industry is not very interested. However, some academic laboratories, including that of James Shorter at the University of Pennsylvania, are looking for small molecules that inhibit self-templating. They take advantage of their extensive knowledge of yeast prion proteins (of which more later in the book) and use yeast prions as a model. Small molecules that interfere with self-templating of tau have been identified in this way. At least one of them is being evaluated in a clinical trial.

Besides blocking self-templating, scientists are also trying to boost protein quality control. The goal is to help neurons to degrade, or to refold, misfolded prion proteins. Hopefully these approaches will lead to clinical trials for Alzheimer's disease treatments in the not-too-distant future.

Intervening at the level of the axonal transport of Abeta and tau aggregates requires that we first learn more about the molecular mechanisms of these events. We know that prion proteins in general can be transported long distances inside axons. We need to know more about this transport. How do Abeta prions enter neurons? Are there specific receptors that could be blocked? How do they exit axons and transfer to more neurons? Is it rapidly, across the very narrow space of synapses? Or do they spend time diffusing outside neurons, time during which they could be reached by drugs? And what about local diffusion of Abeta seeds from the amyloid plaques that are outside neurons? Does that play a part in the neuron-to-neuron spread? All these questions and many more need answers to orient the design of drugs to interfere with spread.

Are Alzheimer's and Parkinson's Diseases Infectious?

The answer is a resounding NO. The spread of Abeta, tau, and alpha-synuclein prions inside the brain *mimics* an infection by a microbe, such as a virus. The fibrils spread from neuron to neuron and "multiply" by self-templating. But there is no spread of the prions between individuals. These diseases are not "contagious." One does not "catch" Alzheimer's or Parkinson's disease by contact with patients, including with their blood or secretions. The epidemiology is very clear on this point. The same is true for all the other human diseases known to be associated with prion proteins. The only known cases of transmission were of Creutzfeldt-Jakob disease accidentally after neurosurgery and after treatment with growth hormone extracted from human pituitary glands.

Why are human prion diseases not contagious? The PrP prion diseases known to be contagious are all transmitted by eating contaminated food. In animals, scrapie of sheep and "mad cow" disease are transmitted through the digestive tract. For scrapie, by eating contaminated grass and also placenta and other birth products, which sheep eat after ewes deliver. In humans, transmission by food was the case for kuru and most likely for the variant form of Creutzfeldt-Jakob disease that was caused by eating meat from "mad cows." If all prion proteins behave like PrP, the most likely explanation for the absence of interhuman transmission of the prion proteins causing Parkinson's, Alzheimer's, and some other neurodegenerative diseases is simply that we are not exposed to them in our food. Of course we do not even know if transmission through the digestive tract is possible for any of these prion proteins. It could be that digestive transmission is limited to PrP, and therefore to spongiform encephalopathies.

To Recap

Alzheimer's disease is a progressive dementia that affects first short-term and then long-term memory.

Its characteristic brain lesions, described by Alzheimer at the turn of the twentieth century, are called *amyloid plaques* and *neurofibrillary tangles*. Amyloid plaques are outside neurons and are made of aggregated proteins, the principal one being *Abeta*. Neurofibrillary tangles are inside neurons and are also made of aggregated proteins, mainly of a protein called *tau*.

Eva and Heiko Braak, two German neuropathologists, described the series of steps in the progression of the brain lesions. The spread of tau aggregates (neurofibrillary tangles) correlates well with the progression of the dementia. That of Abeta aggregates (amyloid plaques) does not. Tau aggregates spread through the brain between areas connected by axons.

Work done over the past twenty years in several laboratories has shown that Abeta and tau are prion proteins. Abeta and tau make typical prion fibrils in amyloid plaques and neurofibrillary tangles, respectively. Abeta and tau fibrils produce *seeds*, which spread the disease. The prion properties of both Abeta and tau have been shown in neurons in culture and in mice.

The progression of the Abeta and tau aggregates can now be studied in the brains of patients using a new in vivo imaging technique called *positron emission tomography* (PET). This is especially useful for monitoring the progress of clinical trials.

Genetics has been very helpful in studying the cause of Alzheimer's disease. Studies of familial cases have demonstrated that Abeta does have a role in the disease. Methods that compare very large numbers of patients and healthy controls have shown that tau is an important risk factor. These studies also demonstrated that inflammation in the brain is a factor in the death of neurons and in dementia.

Abeta aggregates are toxic for neurons. They inhibit *mitochondria,* which are the organelles that transform oxygen into chemical energy inside cells. Tau aggregates are also toxic, but the mechanism of their toxicity is less well defined.

The genetic studies have identified targets for drugs and treatments. Removing Abeta aggregates from amyloid plaques using antibodies has been approved as a treatment, since it significantly delayed the progression of dementia in clinical trials. Removing tau aggregates is still at the stage of preliminary trials.

Understanding how the Abeta and tau prions spread in the brain is essential to find drugs that will stop the progression of the disease. It should be possible to design drugs to block self-templating or to block the various steps of axonal transport and neuron-to-neuron transfer. More research is needed on these topics.

Alzheimer's disease is not an infectious disease. One cannot "catch" Alzheimer's disease by contact with patients or their secretions.

CHAPTER 6

More Human Diseases Caused by Prion Proteins

Different "Strains" of the Tau Prion Cause Different Diseases

In the previous chapter we discussed the role that the tau prion protein plays in Alzheimer's disease. As it turns out, tau aggregates are found in more than twenty other brain diseases, a staggering number, including a dementia called Pick's disease. Some of these tau diseases are quite rare. They are collectively now referred to as *tauopathies*. Given with these observations, neuroscientists were faced with a question that we encountered before for alpha-synuclein: How can the same protein cause twenty different diseases? Are tau aggregates really a cause or simply a consequence, a way for neurons to react against different sorts of injuries?

A remarkable series of experiments, especially those performed in the laboratory of Marc Diamond at Washington University, convincingly demonstrated that each different tauopathy corresponds to a different folding of the tau protein into a prion. These are prion "strains," caused not by DNA mutations but by the shape taken by the protein. Let's describe some of these experiments in detail, as they illustrate a well-conducted, rational approach to the problem.[1]

In an article from the Diamond laboratory, Sanders and collaborators showed that tau fibrils made in the test tube acted as seeds when added to cultured human cells that made the tau protein. The seeds caused the aggregation of the cellular tau and the formation of new fibrils inside the cells. The aggregates could be seen in the cytoplasm of the cells using a fluorescent reagent. This was of course interesting, since it confirmed that tau fibrils could behave as prion proteins.

However, the scientists noticed that the aggregates had different shapes in different cells. Some cells had a single or a few large aggregates. Others had multiple small, punctiform aggregates. The cells with their aggregates were still growing. Therefore, the scientists could isolate single cells containing only one form of aggregate and grow them separately. They observed that all the daughter cells of cells with large aggregates had large aggregates, and, conversely, all the daughter cells of cells with small aggregates had small aggregates. The size of the aggregates was maintained during seeding and self-templating of the tau prion protein.

1. D. W. Sanders, S. K. Kaufman, S. L. DeVos, A. M. Sharma, H. Mirbaha, A. Li, S. J. Barker, et al., "Distinct Tau Prion Strains Propagate in Cells and Mice and Define Different Tauopathies," *Neuron* 82 (2014): 1271–88.

When the researchers extracted tau fibrils from cells with large or small aggregates and used them as seeds, they observed that seeds from cells with large aggregates produced only large aggregates, and the seeds from punctiform aggregates produced only punctiform aggregates.

Both results were remarkable, because the tau proteins forming the large or punctiform aggregates had the same sequence of amino acids. Yet, they produced different kinds of aggregates, and the size of the aggregate was transmitted to the progeny during self-templating. In classical genetics, this is what one observes with proteins that differ by one or several amino acids, by one or several mutations. But here no mutations were involved.

Later, the scientists were able to purify the aggregates and showed that the folding—the shape of the tau prion protein—was different for each "strain." They had two strains that differed only in the shape taken by the protein when it folded into a prion.

The work went further. The researchers injected mice with the two tau strains and observed that they developed different diseases. They extracted tau fibrils from the brains of the mice and injected them into new mice. The disease produced in this new round of injections was the same as in the original round. Therefore, the "strain" phenomenon shown in cultured cells was also seen in live animals. Again, with no mutation in the sequence of the protein but with a different folding, they obtained different diseases, which could be transmitted from mouse to mouse.

In subsequent work, scientists in Diamond's laboratory and others worked on the tauopathies, including Pick's disease, for which they could obtain brain samples.[2] They extracted tau fibrils from the brains of deceased patients, and looked at their structure using cryo-electron microscopy. Each disease that they were able to study corresponded to a different folding of the tau protein. When they injected these fibrils into the brains of mice transgenic for the human tau protein gene, they obtained different diseases.

So what determines that tau fibrils will give Alzheimer's disease to one patient and Pick's disease to another? Why does it fold this way or that in different patients? We do not have a complete answer to this question, but there are reasonable hypotheses. There may be local factors. There are many different types of neurons: neurons with different shapes, using different neurotransmitters, involved in many different functions. All of them contain many copies of tau protein. The type of neuron in which the initial misfolding and seeding take place may influence the shape taken by the misfolded tau protein. Another possible scenario involves pure chance, or rather mischance. We have already encountered this problem in the case of alpha-synuclein and the different synucleinopathies. As we saw in chapter 2, the folding of a protein into a prion form is essentially due to chance. Intrinsically disordered proteins, such as tau and the Abeta peptide, fold and unfold in myriad different short-lived, unstable shapes. The chance that a molecule of a given shape becomes stable, presumably by binding to another

2. The familial form of frontotemporal dementia is a tauopathy caused by mutations in the gene coding for the tau proteins. Therefore, this tauopathy is not relevant to the present discussion. The other tauopathies are caused not by mutations but by alternative folding of the tau protein.

molecule with the same shape, is very small. We can hypothesize that the folding that gives rise to the tau prion associated with Alzheimer's disease may occur more frequently than the others, or may have a slightly longer life, increasing the chance that two or more molecules with this same folding will aggregate to form a stable prion seed. Once a seed is formed, the process that makes fibrils, and then large neurofibrillar tangles, is on its way, propagating the folding of the original seed.

These are only hypotheses about the mechanism of strain formation for a single prion protein. Another unsolved question is the association of a given strain with a particular disease. Why do subtle changes in the folding pattern of the prion protein cause totally different diseases? Again, we do not have answers for these questions. They are at the forefront of research on prion diseases. Solving them may not only help in devising treatments but also uncover new and fundamental aspects of the relationship between the structure of proteins and their function, a very exciting area in fundamental biology. We will see in the next chapter that "good" prion proteins are used by cells to perform various functions. Could it be that alternate folding of "good" prions is another level of regulation of the biochemical activity of some of our proteins?

There Are Other Human Diseases Associated with Prion Proteins

Several other neurodegenerative diseases of humans, besides Parkinson's disease, Alzheimer's disease, and the tauopathies, are caused by prion proteins. One rather frequent one is amyotrophic lateral sclerosis (ALS), also known as Lou Gehrig's disease. In ALS at least two proteins, SOD1 and TDP43, behave as prion proteins.

Huntington's disease is another. This is a genetic disease, caused by mutations in the gene coding for a protein called huntingtin. The mutations increase the length of an intrinsically disordered domain of the protein by adding to it a variable number of an amino acid called glutamine. Above a certain critical number, which happens to be thirty-six, the protein tends to misfold into a prion form, which forms aggregates inside neurons. Thirty-six glutamines is a threshold. But some patients have more. The greater the number of glutamines in the intrinsically disordered domain, the earlier the onset of disease. This is because the chance of misfolding into a prion increases as the disordered domain of the protein gets longer.

So far, all the prion diseases that we have encountered are diseases of the brain. Why would there be such an association? It could be that because brain cells, especially neurons, do not regenerate efficiently, the loss of neurons leads to disease rather rapidly. But the reason might also be purely historical. The prion field emerged from work on scrapie, kuru, and Creutzfeldt-Jakob disease, all of them brain diseases. Prions and human prion diseases became the field of a small group of neurologists and neuroscientists. Also, the interest in these diseases, and in prions, remained limited until the transmission of Creutzfeldt-Jakob disease by contaminated growth hormone, and the economic disaster caused by bovine spongiform encephalopathy, "mad cow" disease, alerted doctors and politicians to their importance. As a result, new groups of scientists became interested and began looking for prion diseases outside the brain, especially among what are called *amyloidoses*.

Amyloidoses are diseases that have been known for a long time. They affect various organs and are characterized by the presence of deposits of insoluble, aggregated proteins in tissues, which were named *amyloid deposits* by pathologists. As described in chapter 2, "amyloid" simply means that the aggregates show a particular color under the microscope if the tissue sections are stained with dyes such as Congo red and illuminated with polarized light. Many different proteins can form amyloid deposits. What all these deposits have in common is the presence of multiple beta strands stacked on top of one another, making beta sheets—just like in the fibrils of prion proteins. The dyes that stain amyloids recognize and bind to these beta sheets. We have already mentioned that prion proteins form amyloid fibrils. However, not all amyloid protein fibrils are prions capable of self-templating and propagation. It is likely that only a small number of amyloids can self-template.

A common amyloidosis in developed countries is type 2 diabetes. It is linked to both genetic and environmental—lifestyle—factors. It is caused by a deficit in the production of insulin by the pancreas, as well as insulin resistance in the tissues. Back in the 1960s, pathologists reported the presence of amyloid deposits among the cells making insulin in patients' pancreases. The protein forming these amyloid deposits has been identified more recently and is called *islet amyloid polypeptide* (IAPP). The aggregated form of IAPP appears to be toxic for the pancreatic islet beta cells, the cells that make insulin. In 2017 the laboratory of Claudio Soto, at the University of Texas in Houston, published an article that showed convincingly that IAPP in amyloid fibrils is a prion protein. They were able to cause diabetes in several animal models by injecting them with IAPP aggregates extracted from diabetic pancreases, or with fibrils made in vitro from purified IAPP. Thus, the slow progressive spread of lesions within the pancreas during type 2 diabetes appears to be due to the "infectivity" of aggregated prions of IAPP. Type 2 diabetes is only one example of a prion amyloid outside the brain. The hunt is on for more. It is very likely that others will be reported in the future.

To Recap

Besides Alzheimer's disease, the tau prion protein is also involved in twenty different *tauopathies*. With one exception, they are not caused by mutations in the tau gene; they differ by the folding of the tau protein into several different shapes.

What causes the multiple different folding patterns is not known. Hypotheses range from pure chance to the chemical environment in the type of neuron in which misfolding and self-templating first take place.

Prion proteins are involved in several other neurodegenerative diseases besides Alzheimer's and Parkinson's diseases and the tauopathies. *Amyotrophic lateral sclerosis* (Lou Gehrig's disease) is one.

Huntington's disease is a genetic disease caused by the lengthening of an intrinsically disordered domain of the huntingtin protein. Above thirty-six glutamine amino acids, this domain folds into a prion. The greater the number of glutamines, the earlier the disease begins.

Not all human prion protein diseases are diseases of the brain. Type 2 diabetes, a common disease, is also associated with a prion protein, which spreads lesions in the pancreas.

CHAPTER 7

A "Good Prion" Protein Needed for Memory

The diseases that led to the discovery of prions—scrapie, kuru, Creutzfeldt-Jakob disease, and other spongiform encephalopathies—are diseases of the brain that strongly affect memory. It is ironic that the first "good prion" to be discovered in animals—a prion protein that instead of causing disease is needed for the normal function of the cell—was a neuron protein involved in long-term memory. Others useful prion proteins have subsequently been discovered, and one can safely predict that more will be found in the future. In fact, self-templating proteins might be one of the many categories of proteins that perform important functions in cells. Only in exceptional cases are some of them responsible for disease, because they are toxic and can spread from cell to cell, and even from individual to individual in the case of some of the spongiform encephalopathies.

A Detour through the Brain and Memory

To study memory, scientists and clinicians need to divide this complex phenomenon into simpler, well-defined categories. For example, one can distinguish between short-term and long-term memory. Short-term memory holds the information you need to perform everyday tasks such as dialing a phone number that you just read or comparing the price of organic versus ordinary strawberries at the supermarket. It also keeps track of where you are: in which room, in which building. This information is quickly forgotten after performing the task; thankfully, since you do not want to overload your brain with such trivia. However, the loss of short-term memory, as in Alzheimer's disease, makes patients unable to carry out daily routine activities, such as getting dressed and making breakfast. It progressively makes them totally dependent on external help.

Long-term memory is the memory of events and information that has been stored in your brain for long periods, for as long as you have been alive in many cases. The amount of such information is enormous. Think about your memories of your childhood, or how long you have remembered your name, your birthday, and how much is nine times seven. In great part, long-term memory shapes our personality, makes us who we are. Losing it is devastating; by erasing the past it robs us of our identity.

Long-term memory can be subdivided into several categories. *Episodic* memory, as its name implies, is the memory of events. Where it happened (space), when it happened (time), and everything else connected to that event—in brief, it is everything we remember about our life since childhood. *Semantic* memory is the memory of words, numbers, concepts, and so on. It is the memory that allows us to perform mental tasks and to communicate with others using language. Another category is *procedural* memory. This is the memory of skills, such as riding a bicycle, that you have learned but don't need to consciously retrieve to perform—a memory that is automatically recalled when you need it.

We now know beyond any doubt that memory is a product of the brain, and that in the brain it involves neurons and their connections. Remembering something goes in stages. The first one is *acquisition*. This is a complex series of interactions between sensory neurons for vision, hearing, touch, and so on, and neurons in various other areas of the brain. Among these, an area called the *hippocampus* is a major center of activity for memory. The next step is *consolidation*. This important step makes the difference between short-term and long-term memory. There is no consolidation for short-term memory; it is erased. To create long-term memory, new connections between neurons in the hippocampus and several other parts of the brain must be established. Consolidation, which takes place mainly during sleep, creates a network of neurons across most of the brain. This network is sometimes called an *engram*. Finally, there is *retrieval*. This is what happens when a cue, a voluntary effort to remember something, or some external event, reactivates the network, the engram, and makes us conscious again of something that has been hidden in our brain, sometimes for many decades.

This scheme brings to mind a host of questions. Which neurons are involved in these various steps, including in the case of long-term memory? What is the difference between a short-lived network of neurons during short-term memory and a long-lived one after consolidation? What directs some memories to consolidation and others to erasure? Which molecules are involved in the neurons at each of these stages? These questions, and many others, have puzzled neuroscientists for a long time, and although we are still a long way from having complete answers, remarkable advances have been made over recent decades. The mechanisms of memory is one of the most active fields of neuroscience.

Memory is a neuron problem. Therefore, let's quickly review what we learned in chapter 3. An excited neuron *fires* a short electrical nerve impulse, an *action potential,* which travels down its *axon*. The current is weak but can be measured with electrodes and amplifiers. Neurons communicate with each other at the level of *synapses,* the points of contact between two neurons. At a synapse, the *nerve impulse,* the information, travels in only one direction, from the axon of an upstream neuron to a *dendrite* of a downstream neuron. The action potential coming from the axon of the upstream neuron will trigger a new action potential in the downstream neuron only under certain conditions. The synapse regulates the transfer of information and can make it stronger or weaker. Some synapses, from neurons called *inhibitory neurons,* prevent the downstream neuron from firing action potentials. Synapses are possibly the most important components in the transfer of information across the brain. They have been compared to transistors in a computer, but they are adjustable transistors, they regulate the transfer of impulses between two neurons. They are flexible.

At the synapse, part of the membrane of an axon terminal of the upstream (*presynaptic*) neuron comes into close contact with part of the membrane of a dendrite of the downstream (*postsynaptic*) neuron. Synapses can be observed under electron microscopes (see figure 3.3). On the dendrite side, they appear as globular structures with a stem that connects them to the dendrite. On the axon side they also appear globular but are filled with small vesicles, the *synaptic vesicles* (of which more later). Synapses are complex structures, containing many different proteins. They also contain messenger RNAs coding for synaptic proteins.

The human brain contains approximately eighty-six billion neurons, which communicate through close to quadrillion (10^{15}, or one followed by fifteen zeroes) synapses. Those are astronomical numbers. It is difficult, or even impossible, for us to construct a mental image of a quadrillion objects. It is too large. For example, one hundred billion is the number of stars in our galaxy, the Milky Way. The number of synapses in our brain is equivalent to the number of stars in ten thousand Milky Ways! This gives you an idea of the complexity of the brains of higher animals and humans.

The synapses involved in memory are *chemical synapses*. When the action potential in the axon of the upstream neuron reaches the synapse it releases molecules of a *neurotransmitter*, a chemical substance, into the *synaptic cleft*, the space that separates the two neurons. These neurotransmitter molecules are stored in the synaptic vesicles that we can see by electron microscopy (figure 3.3B). The neurotransmitter diffuses into the synaptic cleft and binds to receptors on the other side, on the membrane of the downstream neuron. This binding triggers an electrical current in the dendrite of the downstream neuron. The mechanism by which an incoming action potential causes release of a neurotransmitter and by which binding of the neurotransmitter to the receptor triggers a new action potential is complex. It depends on the nature of the neurotransmitter and of its receptor. It also depends on the physical structure of the synapse. These mechanisms have been studied extensively and are quite well understood. We reviewed some of these mechanisms in chapter 3.

You may wonder why there is such complexity in the communication between neurons. Why wouldn't the action potential, an electrical current, simply jump from one neuron to the next at the synapse? In fact, such *electrical synapses* do exist. They are very fast and are used for reflexes between sensory and motor neurons. But they act always in the same stereotypic fashion; they cannot be adjusted. The complexity of chemical synapses is precisely what makes them able to regulate the flow of information between neurons, a bit like dimmer switches that can be set to give different levels of light in a room. This is essential for the formation of memory. In fact, engineers are trying to imitate this with what they call *synaptic transistors*—transistors that can "remember" or "learn" what they have done previously. Development of synaptic transistors is an active field of computer science.

From the synapse on a dendrite, the electrical current with its potential travels all the way to the cell body of the downstream neuron, where it triggers a cascade of chemical events that may result, again under certain conditions, in sending an action potential down its axon. As we mentioned in chapter 3, in many instances the cell body of the neuron, its *soma*, sums several potentials coming from different synapses. Depending on the result, it may fire an action potential down its axon. In a similar fashion, the neuron that receives this action potential, if one has been fired, may connect to a third one, and that to a fourth, and so on. If you compare the number of neurons and the number of synapses in our brain, you will conclude that each neuron has several thousand synapses. Any given neuron may receive signals from ten thousand other neurons through synapses on its dendrites. Its job is to integrate the information that it receives and "decide" whether to send an action potential down its axon.

Memory and Synapse Remodeling

What do synapses have to do with memory? As already mentioned, memory consists in creating networks of neurons belonging to different parts of the brain, all of them connected by synapses. When making a short-term memory, the synapses involved in the new network are not physically changed, and they return to their previous state more or less quickly after the formation of the network. Short-term memory disappears.

In contrast, in long-term memory, either because the synapses are repeatedly stimulated, as when we learn a new task, or because other parts of the brain, including parts that deal with emotions, send chemical signals to the synapses of the network, the synapses undergo profound changes—they are *remodeled*. They become larger. New synapses may sprout next to the original one. Synaptic remodeling requires the synthesis of many different proteins; therefore, it takes some time.

The physical changes in synapses over time—their enlargement, the sprouting of additional synapses—can be followed in the brains of live mice during acquisition of memory using sophisticated in vivo microscopes. The reality of synaptic remodeling is very well documented. It increases the strength of the connection between the two neurons. It makes the synapses more responsive, more powerful. The transfer of information between the neurons is made more efficient.

The idea that synapses are not static but may change depending on circumstances goes back to the 1940s and the work of Donald Hebb, a famous Canadian psychologist. Eighty years later, following much experimental work undertaken by many groups of scientists, the synapse remodeling model is the most convincing mechanism to explain long-term memory.

Memory and Neuron Networks

Some remarkable work done in the laboratories of Susumu Tonegawa at the Massachusetts Institute of Technology and of Sheena Josselyn at the University of Toronto confirmed this model with elegant experiments performed in mice. These researchers resurrected the "engram" concept that had been proposed at the turn of the twentieth century by Richard Semon to explain long-term memory. In its modern form, an engram is a network of neurons connected through remodeled synapses, which is created during the acquisition and consolidation of long-term memory. The neurons of a given engram may belong to many different parts of the brain. The engram may be silent for long periods of time but is reactivated during retrieval of the associated memory. According to the model, retrieval is due to a cue, such as the taste of tea and of the "petite madeleine" pastry in the famous episode of Proust's masterpiece *In Search of Time Lost*. The cue, which in Proust's case came from sensory neurons, somehow connects these neurons to those of the silent engram and activates retrieval.

To test the engram model, Josselyn, Tonegawa, and their respective collaborators reasoned, first, that if an engram exists, one should be able to observe a network of neurons firing together in the brain during training for a task that the animal will remember. Second, that one should be able to erase that memory by inactivating at least one neuron in the network. Third, that one should be able to retrieve the memory by stimulating one of the neurons of the network in the absence of a natural cue.

All these predictions were confirmed in a series of remarkable experiments (figure 7.1). Using transgenic mice with light-emitting neurons (see chapter 3) the scientists were able to map networks active during learning and during retrieval. Using optogenetics, a technique described in more detail in chapter 3, they silenced a single neuron in the network and observed that the mice lost the memory associated with their training. Again with optogenetics, they could activate a neuron of the network and retrieve the memory in the absence of the normal recall cue. For example, mice trained in a fear-conditioning test to freeze when hearing a white noise could be made to freeze in the absence of the white noise by stimulating one neuron in the engram network. Even more impressively, by stimulating specific neurons with optogenetics the

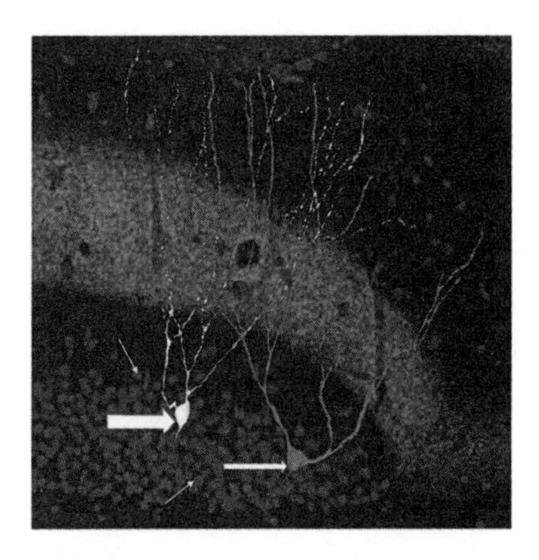

FIGURE 7.1. A neuron involved in an engram for fear conditioning. The mouse had undergone training in a fear-conditioning test that associated pain with a particular noise. Both neurons shown with arrows were engineered to produce a neuron-specific fluorescent dye and a dye specifically turned on when a neuron is firing action potentials. The neuron indicated by large arrow produced both dyes; the one with the medium-size arrow produced only the neuron-specific dye. The neuron indicated by the large arrow, which keeps firing, is part of the engram for the memory of the fear associated with the noise during training. Both the large and middle-size arrows point to the cell bodies of the neurons. The dendrites extend upward from the cell bodies; the axons extend downward. Only the beginning of each axon is visible. The thin arrows point to the nuclei of two neurons in a layer of neurons called the dentate gyrus of the hippocampus, which has a major role in the acquisition and retrieval of memory. From "Memory Engrams: Recalling the Past and Imagining the Future" by Sheena A. Josselyn and Susumu Tonegawa (*Science* 367, no. 6473 [2020]: eaaw432); © 2019 the authors. Originally published in "Engram Cells Retain Memory under Retrograde Amnesia" by T. J. Ryan et al. (*Science* 348, no. 6238 [2015]: 1007–13); © 2015, American Association for the Advancement of Science

researchers were able to introduce a particular memory in the brains of mice that had not been trained. In other words, they could create the memory of an event that never occurred.

The results on networks and engrams and the strong evidence that stable connections between neurons are due to synapse remodeling form the present framework of the mechanism of long-term memory.

About Memory in a Sea Slug

Eric Kandel is a famous neuroscientist who has been studying the mechanisms of memory at Columbia University for many years.[1] Early on, in the 1970s, he decided that the best and possibly only way to solve such a complex problem was to choose an animal model that was as simple as possible. He settled on a sea slug called *Aplysia californica*. Why? The animal has a simple nervous system with only a few thousand neurons, while our brain contains eighty-six billion of them. Its neurons are extremely large—some are visible with the naked eye—which makes them easy to study. Furthermore, the behavior of the animal had been studied, and some reflexes had been characterized by other scientists. One of them was a measurable fear-conditioning test. This test went as follows: when touched on a certain part of its body called the *siphon*, the animal retracts another part of its body, called the *gill*. If the animal is given a slight electric shock to its tail while one touches the

1. Kandel's book *In Search of Memory* is a wonderful introduction to neuroscience in general as well as a fascinating autobiography (New York: W. W. Norton, 2006).

siphon, the retraction of the gill is stronger and lasts longer. The animal has been *sensitized*. If it is tested some days later by touching the siphon, but without the electric shock, it still shows a strong retraction of the gill, and it remains sensitized for several days.

We should pause for a moment and consider the importance of the choice of the model for solving a complex biological problem. Of course, *Aplysia* with its syphon and gill is, in terms of memory complexity, far removed from our brain and the memories of our childhood. But Kandel knew that he needed a much simpler system, with only a few neurons and a behavior that could be quantified. At the time he was searching for a suitable model, the sensitization of *Aplysia* to an electric chock had already been described by others. With the available information in hand, creativity came into play. Kandel and his coworkers realized that sensitization could be used as a form of memory. Because the animal remains sensitized for some time, it must *remember* that touching its gill caused a disagreeable sensation in its tail. Here was a simple nervous system that could *remember* an event. On top of that, this memory could be quantified by measuring the extent of retraction of the gill and the number of days that the animal remained sensitized. This critical choice led to the discovery of the role of synapse remodeling in memory and of a prion protein needed for this process. A good choice indeed.

Having a way of measuring memory was essential. Measurements are the basis of all good experimental science. Scientists progress in their research by making and testing hypotheses, models that they build to explain what they observe. To test a hypothesis, they devise experiments, which most often consist in perturbing their model. For example, in the case of memory and neuron networks, a scientist may reason: "If the hypothesis is correct, then when one stimulates such and such a neuron the animal should remember its training for a longer period of time." The experiment consists in measuring the memory and comparing the results with and without the stimulation of the neuron whose function is being tested. The quality of the measurement—its accuracy, its reproducibility—is critical. Above all, experimental science depends on good quality measurements.

Using *Aplysia* and the fear-conditioning test, Kandel's team made a series of remarkable observations. First, they determined that repeated stimulation of the siphon was needed to create a good, long-lasting memory. With only one or a few stimulations, the memory did not last. This gave them a model for short-term versus long-term memory. Then, they determined which neurons were involved in the memory test. The circuitry turned out to be remarkably simple. A sensory neuron carries the feel of touch from the siphon to the primitive brain, where it connects to a motor neuron, which commands the retraction of the gill. Another sensory neuron carries the feeling of pain caused by the electric chock to the tail. It connects to the sensory neuron that carries the sense of touch. Four more *interneurons* are part of the circuit.

This circuit is the same in all *Aplysia* animals. Moreover, the scientists observed that training in the fear-conditioning test, creating long-term memory, did not change the circuitry. They concluded that acquisition of memory is not due to building new connections between neurons. Instead, by stimulating neurons and recording their responses, they observed that what changes during training is the strength of the connections, of the synapses between sensory and motor neurons. Synapses that are repeatedly stimulated during training become stronger, more efficient at transmitting a signal to the downstream neuron. This important conclusion led to the now well-accepted concept that long-term memory is associated with synapse remodeling. Kandel's experimental work with *Aplysia* provided a wonderful vindication of the model for memory proposed in the 1940s by Donald Hebb. It also won Kandel his Nobel Prize.

Because *Aplysia*'s neurons are very large and robust, the researchers in Kandel's laboratory were able to reconstruct the circuit between sensory and motor neurons in a culture dish with neurons taken from the brain of the animal. They also identified the neurotransmitter used by the synapses between these neurons. Having an in vitro circuit of neurons that could be stimulated with the authentic neurotransmitter made it possible to study the molecular mechanism of synapse remodeling. The main conclusion of a series of elegant experiments was that synaptic remodeling, and long-term memory, requires the synthesis of many different proteins at the level of the synapse.

A Prion Protein for Long-Term Memory

The neurons in the culture dishes of Kandel's laboratory had only a few synapses. In the brain, even in primitive animals, a neuron may have thousand synapses. Synaptic remodeling of a few or even only one of them is needed for the formation of a new engram and the acquisition of memory. This raised a difficult new question for Kandel and his team. How does the neuron *know* which synapse has been stimulated repeatedly during training and requires protein synthesis to be remodeled? Kandel and one of his postdoctoral fellows, Kausik Si, reasoned that the stimulation during training may leave a molecular tag in the synapse that tells the neuron that protein synthesis is required there and not in other synapses. What could the tag be? Fortunately, when Si and Kandel were pondering these questions in the early 2000s, other researchers had already determined a list of proteins present in synapses. Si and Kandel searched the list for a candidate and became interested in a protein called CPEB, because it was known to be needed for protein synthesis.

You may remember from chapter 2 that the blueprint, the *sequence*, of a protein is encoded into the DNA of its gene. To make the protein, the gene is first transcribed into an RNA copy of the DNA, a messenger RNA, or mRNA for short. This mRNA is then translated by ribosomes into the chain of amino acids that makes the protein (see figure 2.3). But before it can be translated into a protein, the mRNA needs to be modified—extended at one of its extremities by a sequence called a *poly(A) tail*. This extension tells ribosomes that the mRNA is ready for translation into a protein. mRNAs remain dormant in the cytoplasm of the cell until a poly(A) tail is added to their extremity. CPEB, Kandel and Si's candidate for the tag, is a protein that adds a poly(A) tail to mRNAs.

Si first determined that CPEB was present in *Aplysia*'s neurons (the list of synaptic proteins had been established for the mouse). He noticed that the protein had an intrinsically disordered domain. At the time, the role of the intrinsically disordered domains of the PrP protein and the misfolding of the protein into a prion, the scrapie agent, were hot topics in the protein field. Might CPEB behave as a prion? Si and Kandel wondered. To test this hypothesis, they teamed up with Susan Lindquist, an expert in yeast prion proteins at the Massachusetts Institute of Technology. They replaced the intrinsically disordered domain of a well-characterized yeast prion by that of CPEB. It made no difference to the yeast. The yeast cells functioned in prion assays just like normal yeasts with their authentic yeast prion protein.

This remarkable result led Si and Kandel to formulate a new hypothesis: that repeated stimulation of the synapse during training turns *Aplysia* CPEB protein into a prion form, and that the prion form was the tag that they had been looking for. As a result of self-templating, the CPEB prion grows into a fibril, which is then too large to move into the neuron and trigger remodeling in other synapses within the neuron. Self-templating ensures that the synapse remains tagged, and that memory persists, despite the normal turnover of proteins. They also hypothesized that it is the prion form of CPEB that adds the poly(A) tail to dormant mRNA, turning on the synthesis of proteins required for synapse remodeling.

CPEB was the first "good prion" discovered in animals. One facet of this example is worth emphasizing. For PrP, alpha-synuclein, and other disease-causing prion proteins, folding into a prion is rare and due to chance, hence the term *misfolding*. In the case of CPEB, folding into its prion form is a normal process that is induced by the activity of the synapse. The process is precise, accurate. The prion form performs a function; it allows the synthesis of the proteins needed for synapse remodeling and memory.

Following this pioneering work, several laboratories, including that of Si, now on the faculty of the Stowers Institute for Medical Research in Kansas City, and that of Krystyna Keleman at the Howard Hughes Medical Institute, pursued the study of CPEB and long-term memory. They turned from *Aplysia* to the fruit fly *Drosophila melanogaster*. *Drosophila* is a classic animal model in genetics and neuroscience, and its nervous system has been studied in detail. Several long-term memory tests have been designed by neuroscientists over the years for this insect. Scientists know how to modify its genes by genetic engineering. It offered experimental advantages that *Aplysia* did not have.

Drosophila's neurons have a CPEB protein, which had been called Orb2 by geneticists. It has an intrinsically disordered domain. With *Drosophila* it became possible to manipulate the *Orb2* gene by genetic engineering. Using these molecular tools, the researchers showed that preventing the formation of the prion form of Orb2 protein impaired the acquisition and the retrieval of long-term memory. Conversely, facilitating Orb2 aggregation into a prion protein allowed the flies to form memory even with minimal training.

These were impressive results, but there was still no direct evidence that Orb2 could turn into a prion in vivo, in the fly, rather than just in yeasts and in test tubes. This changed when, in 2020, Si and his collaborators published an article in the journal *Science* in which they described the structure of the Orb2 prion purified from the brain of adult *Drosophila* flies.[2]

2. The article describes a tour de force in biology: R. Hervas, M. J. Rau, Y. Park, W. Zhang, A. G. Murzin, J.A.J. Fitzpatrick, S.H.W. Scheres, and K. Si, "Cryo-EM

They had to use two million fly brains to obtain enough protein to conduct their biochemical experiments! In adult flies, which had been using long-term memory to navigate their environment, the Orb2 protein was present in its soluble form but also in prion-like aggregates. In contrast, only the soluble, nonaggregated Orb2 was found in embryos, whose neurons had not acquired memory. The structure of the aggregated fibrils from adult brains was determined by Si and his collaborators at near-atomic resolution. It had all the characteristics of prion protein fibrils. These fibrils could be broken into seeds that initiated a self-templating reaction with soluble Orb2. In brief, the researchers showed convincingly that Orb2 behaves as a prion protein in the brains of adult flies. Furthermore, they showed that the prion form of Orb2 does indeed add a poly(A) tail to messenger RNAs, whereas the soluble form does not.

The CPEB Prion Protein and Memory in Mice

Sea slugs and fruit flies have simple brains with a rather limited number of neuron networks. It is this simplicity that makes them powerful models. But, of course, they are very distant from humans, and at some point one wishes to study memory in a more relevant system, in a mammal. The most widely use mammal for these studies is the laboratory mouse. Its brain is analogous to the human brain in many respects. It is a bit like a simplified version of our brain.

Structure of a Neuronal Functional Amyloid Implicated in Memory Persistence in *Drosophila*," *Science* 367, no. 6483 (2020): 1230–34.

Many memory tests have been designed for the laboratory mouse. One of them is another example of fear conditioning. When a mouse is introduced into a new cage, its normal behavior is to move around, exploring its new environment. In the test, while the mouse is still exploring its new cage, it is subjected to a gentle white noise and to a slightly painful electric shock in its paws. The electric shock makes the mouse "freeze"; it stops exploring and adopts a typical hunched posture. Noise and electric shock are repeated a few times, then stopped. The mouse is left alone for periods of time that can range from hours to days. After rest, the mouse is moved to a new cage, which it starts exploring. It is then subjected to the same white noise, but without the shock. If it remembers the association between noise and pain, it freezes even though there is no electric shock. If it has forgotten it does not freeze. The number of days that the mouse remembers the association is used as a measure of memory. There are several other memory tests used routinely with mice. Some address more elaborate forms of memory; for example, spatial memory, remembering where objects are in space.

A widely used test for spatial memory is called the Morris water maze. The mouse is introduced into a circular pool filled with opaque water. There are drawings above water level on the sides of the pool, which the mouse can use to orient itself. There is also a small invisible platform in a particular place in the pool, just under the water level. The animal explores the pool by swimming at random until it hits the platform and climbs on it. The second time the animal is placed in the same pool it finds the platform more quickly. It has located the platform with respect to the position of the drawings around the pool. As training continues, it finds the platform more and more quickly. If the animal is returned to the pool one week or longer after training, it still quickly finds the platform. It remembers where the platform is located based on its position with respect to the spatial clues. The speed with which the animal learns the position of the platform and the length of time it remembers it are used as measures of memory.

With mice, one can use genetic engineering to inactivate a gene, to modify the way it is regulated, and so on. Kandel and his collaborators took advantage of these molecular techniques and of the memory tests just described. In a series of experiments in several respects close to those done with *Drosophila*, they showed that long-term memory is severely impaired when the mouse CPEB gene is inactivated.

A sea slug, the fruit fly, mice: a role for the CPEB prion protein in memory appears to be widespread in the animal kingdom. What about humans? We do not have experimental data at present, but it would be very surprising if humans were an exception.

CPEB Is Not the Only "Good" Prion

Other useful prion proteins besides CPEB have been identified and studied. That more will be found in the future is an easy bet. We are only at the beginning of the story of the prion proteins used by cells to perform various functions.

What all these proteins have in common is self-templating (chapter 2). They fold in a particular way (*misfold*, for the bad ones) that can cause other copies of the same protein to acquire the same folding. When there are multiple copies of the protein they can form an aggregate, often in the shape of a fibril, a long pile of the prion-folded proteins. The prions responsible for diseases—PrP of scrapie, alpha-synuclein of Parkinson's disease, and so on—make *seeds,* small pieces of the fibril, which can enter new cells and propagate the misfolding across an organ such as the brain. This is the mechanism by which these diseases are "infectious." In contrast, most useful prion proteins discovered so far are not infectious. They do not spread from cell to cell. The prion form of CPEB and its aggregates remain where they were formed, in a synapse. They do not even diffuse within the neuron. Another fundamental difference is that the misfolding of the prions of scrapie, Parkinson's disease, and the like, is due to chance. It occurs rarely and can be erased by the quality control mechanisms of the cell. In contrast, for CPEB and other useful prions, folding into a self-templating prion form is induced by signals coming from outside the cell. The whole process is precise, having been refined over millions of years of evolution. In this book, for the sake of simplicity, I have decided to call all these proteins *prion proteins*. But we need to keep in mind the fundamental differences between those associated with disease and the ones performing useful, normal functions.

Among the latter, two called MAVS and ASC are essential in the fight again infection.[3] They are part of what is called *innate immunity*—a series of rapid responses that start shortly after

3. A review by Xin Cai and colleagues presents a good summary of the role of prion proteins in the fight against viruses: X. Cai, H. Xu, and Z. J. Chen, "Prion-Like Polymerization in Immunity and Inflammation," *Cold Spring Harbor Perspectives in Biology* 9 (2017): a023580.

the detection of a microbe and are aimed at limiting its spread. We will briefly consider the case of MAVS.

Cells in general, including neurons, are equipped with molecules that detect the presence of viruses in their cytoplasm. This detection is followed by a cascade of reactions between different proteins, which culminates in the production of proteins called *interferons*. As the name suggests, interferons interfere with the multiplication of the virus inside the cell. But interferons are also secreted by the cell. They bind to receptors on neighboring cells and trigger reactions in these cells that make them resistant to the virus. Not only do interferons limit the multiplication of the virus in the infected cell, they also make the surrounding cells resistant. This is a very powerful way of limiting the damage due to viruses; it has a major role in keeping us healthy. Patients with mutations that prevent the production, or the action, of interferons are extremely susceptible to many different viruses.

The MAVS protein is one that takes part in the cascade of interactions that results in the production of interferons. The detection of a virus inside the cells turns the MAVS protein into a prion protein, in a similar way to the repeated stimulation of a synapse turning CPEB into a prion protein. The initial MAVS prion turns all other copies of the MAVS protein in the cell into prions, and they form a large aggregate. It is the MAVS aggregate that triggers the production of interferons. For the cell, and the organism as a whole, the rapid formation of the MAVS prion aggregate works like a trigger. It causes the interferon response to be almost immediately at its maximum level, rather than proceeding incrementally. The prion mechanism of self-templating provides the organism with an important weapon in the race to halt viral multiplication and ensure the survival of the host.

Prion proteins were discovered because some of them cause diseases that look like infections by a classical microbe. We have learned more recently that prion proteins are widespread and play important functions in living organisms. CPEB and MAVS are only two examples of "good" prions. Several others have been discovered in animals. Many exist in yeasts. They have also been found in bacteria and even in some viruses. All this indicates that the prion phenomenon has been associated with life for a very long time. It may even have been at the origin of life on the planet, as we will discuss further in the next chapter.

To Recap

Some self-templating prion proteins play essential functions in cells that harbor them. They are the "good," useful prions. The first one discovered was a neuron protein involved in long-term memory.

Memory can be divided into *short-term* and *long-term* memory. Memory *acquisition* and *consolidation* of long-term memory result from the formation of networks of connected neurons. In short-term memory, the network is unstable and disappears soon after acquisition. In long-term memory, in contrast, the network is made stable and long-lasting by *synaptic remodeling*. Long-term memory and synaptic remodeling require the synthesis of new proteins.

The neuron network hypothesis for memory has been independently confirmed by Sheena Josselyn and Susumu Tonegawa using laboratory mice. They used a revolutionary technique called *optogenetics*. Optogenetics uses light from lasers to make neurons fire *action potentials*, or to prevent them from doing so. With optogenetics the experimenter can control the activity of a given neuron or group of neurons.

Neurons implicated in a memory network have thousands of synapses. For the acquisition of a given memory, only one or a few of these synapses need remodeling. How does the neuron know which synapse needs remodeling? What does stimulation during acquisition do to a synapse to tag it for remodeling?

The laboratory of Eric Kandel approached these questions using a sea slug called *Aplysia californica*, which has a very simple brain. They showed that a protein called CPEB is turned into a prion form by repeated stimulation of a synapse involved in long-term memory. Prion self-templating ensures that the tagging of the synapse for remodeling persists. The prion form of the protein stimulates protein synthesis. The same protein with the same role in long-term memory has been discovered and studied in the fruit fly and in mice. It would be very surprising if humans were an exception.

"Good prions" play essential roles in the immune response to infectious agents. The MAVS prion protein is involved in the production of *interferon*, an essential component of the defense against viruses. Several other good prions have been discovered, including in yeasts.

CHAPTER 8

Where Do Prion Proteins Come From?

Yeast Genetics Challenges Mendel's Laws of Heredity

Geneticists study the way *traits,* also called *characters*—such as eye color in humans, the shape of antennae in flies, or the color of flowers in plants—are transmitted to progeny in species that reproduce sexually. The laws of genetics were discovered in the mid-nineteenth century by an Augustinian friar called Gregor Mendel, who was crossing different varieties of peas and observing how the color and shape of the flowers were transmitted to the next generation. He was collecting pollen from one variety of peas and using it to inseminate another variety. After harvesting the seeds produced by such a cross and sowing them, he counted the number of resulting plants that showed the trait of each parent. From many such crosses, he discovered laws that predicted the proportion of offspring that inherited a given trait from one parent or from the other. Mendel's laws turned out to apply to all species that reproduce sexually, including humans. It would be almost a century before his findings were explained through the discovery that genes are made of DNA and through deciphering the mechanisms by which DNA is replicated and distributed among daughter cells during cell division.

All the cells of an individual contain two copies of each of the individual's genes. All cells, that is, except the gametes, the sperm cells and the oocytes, or eggs. When those are made, they acquire only one copy of each gene. Fertilization, the fusion of oocyte and sperm, reconstitutes a cell with two copies, one from each parent. For each offspring, a given trait may be that of just one parent or just the other parent, or something in between, depending on the trait. Some traits are said to be *dominant*, where only one copy of the gene is sufficient to produce that trait. Others are *recessive*, where it is necessary to have both copies of the gene to confer the same trait. The laws discovered by Mendel predict the proportion of offspring that will inherit a dominant or recessive trait in various kind of crosses.

Yeasts are simple organisms, of the size of bacteria. Like bacteria they do not form complex organs; they live as separate, identical cells. However, they are very different from bacteria. Their chromosomes, with their DNA and genes, are separated from the rest of the cell by a membrane, forming a nucleus. They belong to the category of organisms called *eukaryotes*, organisms whose cells have a nucleus. Besides having a nucleus, their cytoplasm is organized into many compartments, having all the different organelles, including mitochondria, found in the cells of plants and animals (see figure 2.2). During their life cycle, yeasts go through different phases, during which cells contain either two copies or only one copy of each gene. Two yeast cells with only one copy of each gene may fuse to give a new yeast with two copies, very much like sexual reproduction in animals and plants. Yeasts are the simplest available organisms that, at the molecular level, function very much like human cells. What we learn from studying yeasts is directly relevant to what happens in our neurons. Because they divide rapidly and can be grown in artificial media in the laboratory, yeasts are very popular with biologists, in particular geneticists.

Since the 1950s, yeast and mushroom geneticists had been puzzled by crosses that did not follow Mendel's laws of inheritance. In these cases, traits were inherited by all of the offspring, when Mendel's laws predicted that they should be inherited by only a fraction. Could molecular biologists have missed something in the way DNA is replicated and distributed among offspring? In 1994, Reed Wickner at the US National Institutes of Health suggested an answer: these anomalies could be explained if the trait is caused by proteins behaving like prions, rather than by genes made of DNA. Why? When yeasts from two different strains are crossed, they fuse, then replicate their DNA and divide. This is very much like what happens in animals after an egg is fertilized by a sperm cell. Wickner reasoned as follows: During fusion, the proteins from both parents get mixed before being distributed between the two daughter cells. If a trait of one of the two parents is due to a prion protein, both daughter cells will receive some prion. Because prions are self-templating molecules, the prion protein, which comes from only one parent, will turn all the copies of that protein into their prion form in both daughter cells. All the progeny will exhibit the trait, even though it comes from only one parent. If correct, Wickner's hypothesis would describe a new mechanism of genetic inheritance, different from classical Mendelian genetics. In his 1994 article, Wickner went as far as using the term "protein-based inheritance" to describe this process. At the time it was a heretical hypothesis.

It turned out that Wickner was right. Soon after the 1994 article, he and his colleagues reported experiments showing that prion proteins are indeed responsible for the non-Mendelian traits observed in yeast. The conclusion was shocking but inescapable: DNA is not the only molecule responsible for heredity. In fact, protein-based inheritance is not even a rarity. It has now been uncovered in all domains of life, even in some viruses. The discovery of yeast prions created quite a stir among geneticists.

Prion Proteins Are Ancient

Finding prion proteins in simple organisms such as yeasts as well as in complex ones like humans suggests that they are widespread in living organisms and have been associated with life on the planet for a very long time. One way of testing this hypothesis is by comparing the genomes of organisms. This is now possible thanks to fast methods of DNA sequencing and powerful statistical programs to align and compare sequences.

Most prion proteins form amyloid fibrils with their stacks of beta strands forming beta sheets (chapter 2). Not all amino acids can make beta strands and beta sheets. By comparing the sequences of amino acids in beta strands of known proteins it is possible to determine what they have in common and to ascertain the rules that govern their formation. Furthermore, in proteins that make amyloids, the stable beta sheets are usually repeated along the protein chain, separated by loops (see chapter 2). One way of looking for putative amyloid proteins is to look for the presence of these repeated beta-strand sequences in the genes of all the proteins in an organism. The results of this type of analysis suggest that genes coding for amyloid-forming proteins are quite common in the tree of life.

DNA sequence analysis is only one step in finding genes coding for candidate prion proteins. Proteins likely to form amyloids need to be tested to see if they function like prions. A useful test, the one used by Si, Kandel, and Lindquist in the case of CPEB (see chapter 7), takes advantage of assays developed for yeast prions. Using genetic engineering, a well-characterized yeast prion gene is replaced in the yeast chromosome by the gene of the candidate prion (in practice, only the "prion domains" of the yeast and candidate are exchanged; see chapters 2 and 7). The engineered yeast is tested for yeast prion activity. If the test is positive, the candidate is very likely to function as a prion in its original host. Such gene swapping experiments between a yeast prion and a candidate protein are now standard tests for prion activity. Indeed, to function in the yeast a candidate protein must be self-templating. It must be able to confer its folding on other copies of the same protein.

Eventually, a combination of DNA sequence analysis and functional tests has shown that prion proteins are widespread among organisms. Even viruses and archaea, a class of microorganisms different from bacteria, have prion proteins.[1] It appears that prion proteins may have been associated with life since the very beginning. More about this at the end of this chapter.

Prion Proteins and Their Role in Adaptation to the Environment

Living organisms depend on their environment to survive and reproduce. The environment provides them with food from which they derive energy and many essential chemicals. Changes in the environment can create major challenges that exert strong selection pressure on living organisms, with survival of only the most able to reproduce within the new environment. The effect of selection pressure on genes is well known. DNA genomes are faithfully copied with each cell division, but there are some rare errors of replication. These errors can cause mutations, changes in genes that alter the function of the protein encoded by the mutated gene. If the mutation gives the organism an advantage for life in the new environment, the mutant organism will thrive better than the nonmutants, and with time mutant organisms will dominate the population. This is the classical Darwinian selection responsible for the evolution of species over billions of years.

1. T. Zajkowski, M. D. Lee, S. S. Mondal, A. Carbajal, R. Dec, P. D. Brennock, R. W. Piast, et al., "The Hunt for Ancient Prions: Archeal Prion-Like Domains Form Amyloid-Based Epigenetic Elements," *Molecular Biology and Evolution* 38 (2021): 2099–103.

Even within a species, such as humans, there are examples of such selection by environmental pressure. Low oxygen at high altitude has selected for populations in Tibet and the Andes with mutations in genes involved in cell respiration. Infectious diseases have exerted strong selection pressure on the evolution of humans. Several African populations have mutations that give them a certain degree of resistance to malaria. On the other hand, populations in the Americas, which had never been exposed to measles virus and so never been subject to selection for resistant individuals, were devastated by the arrival of Europeans, who brought the virus with them. By now American Indians are as resistant or susceptible to measles as Europeans are, because mutations that determine resistance have been selected for over several centuries (and to some extent because of intermarriage with Europeans).

However, the rate of mutation, of errors made during replication of DNA, is very low. Darwinian evolution is a slow process, acting over very long periods of time. It cannot help organisms survive constant rapid changes in their environment. And when the environment reverts to its previous state, mutations need to revert, another slow process.

But organisms are equipped with some faster mechanisms to deal with rapid changes in their environment. One consists in having a collection of genes that can deal with different environments and a way of turning these genes on or off according to which one is needed. This classical model of gene regulation was discovered in bacteria by François Jacob and Jacques Monod, at the Institut Pasteur in Paris. It is relatively fast and reversible and, with innumerable variations, used by all organisms, not just bacteria.

Another mechanism, also proposed by Jacob and Monod, is called *allostery*. This is where the same protein is able to change its function by changing its shape, its folding, after binding to a specific molecule. Crucially, this refolding occurs in response to a change in the environment. This very much resembles the way "good" prion proteins function—for example, CPEB, which can fold into a prion form during acquisition of memory. However, the big difference between allosteric proteins and prion proteins is that the latter are self-templating, while allosteric proteins are not. Prion proteins spread their new folding to other copies of the same protein, and some of them spread from cell to cell, spreading the new function in the cell population. Allosteric proteins do not do any of that. Nevertheless, one may consider prion proteins as a special case of allosteric proteins. This could be another example of the way evolution proceeds by reusing and modifying existing mechanisms to constantly give new survival advantages to organisms. François Jacob used to call evolution a messy, but successful, tinkering.

Prion proteins are used by some yeasts and mushrooms to adapt to a change in food source in their environment.[2] If a food source disappears, has been exhausted, the protein is induced to refold into an alternative shape, which may help the cell use a different nutrient. This new folding also turns the protein into a prion, which quickly turns all copies of the

2. For an article that demonstrates the adaptative role of yeast prion proteins in changes of environment: R. Halfmann, D. F. Jarosz, S. K. Jones, A. Chang, A. K. Lancaster, and S. Lindquist, "Prions Are a Common Mechanism for Phenotypic Inheritance in Wild Yeasts," *Nature* 482 (2012): 363–68.

protein into its alternative prion form. This is an efficient, fast mechanism to adapt the cell to a new food source. Furthermore, when the yeast divides, all daughter cells inherit the prion and are immediately adapted to the new environment—a great advantage at the population level.

You may wonder what happens if the new nutrient disappears and the old food source comes back. Remarkably, the change of folding that makes the prion is reversible, and the cells can return to their previous state. The mechanism of reversion is not completely understood. It involves protein quality control, the same mechanism that we discussed earlier in this and in previous chapters. On the one hand, protein quality control fragments prion fibrils and creates seeds, which propagate the prion. On the other hand, it may succeed in completely degrading the prion fibrils, thereby "curing" the cell. There is a game of chance going on here. In a yeast population there is always a small number of "cured" individuals. They may not survive for long, but, as the yeasts divide, they are constantly replenished. If the environment reverts to the original conditions, they take over—the population is saved. Protein quality control plays a complicated, balanced game in the life of prion proteins.

Prions Causing Disease Are the Exception

Where do prions responsible for diseases fit in this grand scheme? They appear to be rare, but dreadful, accidents. Like everything in life, things sometimes malfunction. Prion protein diseases are due to misfolding: abnormal folding of an intrinsically disordered protein that, by chance, creates a self-templating protein, a protein that propagates its misfolded shape to more copies of the same protein, a misfolded protein that has evaded protein quality control. A misfolded protein that can invade cells and cause misfolding of the same protein in these new cells. An "infectious" misfolded protein.

Besides the fact that they cause diseases, there are clear differences between the "bad" and the "good" prion proteins. The good ones studied so far don't spread from cell to cell, except to daughter cells after cell division. A fortiori they don't spread from individual to individual. They are part of a mechanism of regulation that must remain specific to a given cell. For example, in memory it is essential that the CPEB prion protein be active not only in a given neuron but *in a given synapse* of this particular neuron. Spreading would ruin everything. The folding of proteins into good prions is induced by external stimuli such as changes in food source for yeast or repeated firing at a synapse in the case of memory. This induction is precise and accurate, having been refined by millions of years of evolution.

In contrast, the folding into dangerous prions is due to chance. It may be a very rare event, as in the case of Creutzfeldt-Jakob disease, which affects only one individual in a million every year, or more common as in the unfortunate case of Abeta and tau in Alzheimer's disease. Even in these dreadful diseases, the misfolding into a prion and spread throughout the brain are inefficient. That is why neurodegenerative diseases with prion proteins are diseases of old age. The longer we live, the greater the chance that the misfolding occurs somewhere in the brain, that protein quality control fails to cope with it, and that seeding in more cells spreads the phenomenon.

Prion Proteins and the Origins of Life

Finally, some scientists wonder if prion proteins might have played a decisive role in the appearance of life on the planet.[3] Life requires a mechanism of self-propagation, of transfer of information between generations, as well as a possibility of evolution into more and more fit and complex forms. Life appeared on the planet around four billion years ago. It may have started in a "soup" of various water-soluble small molecules that assembled spontaneously at high temperature and high salt concentration, most likely in hydrothermal vents, the fissures in the ocean crust, where water is in contact with magma. For a long time, the preferred hypothetical scenario has been that life on the planet started with an RNA world, and many scientists still consider this the most likely hypothesis. According to the RNA-first hypothesis, short RNA molecules may have formed spontaneously in hydrothermal vents. RNA is an attractive "first" molecule because it can encode genetic information, just like DNA, and RNA molecules can function as enzymes—they can catalyze complex chemical reactions.

3. The hypothesis that life on earth began with prion-like proteins has been proposed and discussed by several authors. Among them: J. Greenwald, W. Kwiatkowski, and R. Riek, "Peptide Amyloids in the Origin of Life," *Journal of Molecular Biology* 430 (2018): 3735–50; O. Lupi, P. Dadalti, E. Cruz, and P. R. Sanberg, "Are Prions Related to the Emergence of Early Life?," *Medical Hypotheses* 67 (2006): 1027–33; C.P.J. Maury, "Amyloid and the Origin of Life: Self-Replicating Catalytic Amyloids as Prebiotic Informational and Protometabolic Entities," *Cellular and Molecular Life Sciences* 75 (2018): 1499–1507; S. Jheeta, E. Chatzitheodoridis, K. Devine, and J. Block, "The Way Forward for the Origin of Life: Prions and Prion-Like Molecules First Hypothesis," *Life* 11 (2021): 872.

However, there is a difficulty with the RNA-first hypothesis: self-propagation. Of course, the RNA genomes of present-day RNA viruses replicate, propagate genetic information. But they do it with the help of complex micromachines made of proteins, called *RNA polymerases*. Obviously, RNA polymerases were not present in an RNA-first, pre-life world, in the probiotic original soup. The self-assembly of copies of original RNA molecules poses serious chemical problems for scientists.

An alternative to the primitive RNA world would be a primitive prion world. Some amino acids, identical to or resembling those found in present-day proteins, may have formed in hydrothermal vents, and may have assembled in chains in which they were linked by the chemical bond that links amino acids in present-day proteins: the so-called *peptide bond*. There is experimental evidence in favor of this hypothesis. Scientists have reproduced the conditions found in hydrothermal vents in the laboratory and observed the spontaneous formation of a few different kinds of amino acids and their assembly into short chains, called *peptides*. Furthermore, under these prebiotic conditions, some of the peptides assembled spontaneously into very stable fibrils, because of the formation of beta strands and beta sheets. In other words, the scientists observed the spontaneous formation of amyloid substances under conditions that imitated those found in hydrothermal vents at the bottom of the sea. Those fibrils, like all amyloids, are extremely stable even in harsh conditions. However, they can fragment and function as seeds that propagate the same structure, the same information, to future generations. Remarkably, some amyloid fibrils made in the laboratory under conditions mimicking hydrothermal vents can function as enzymes, as catalysts for chemical reactions that are essential for creating more complex proteins. This primitive period of formation of amyloids, some with prion or enzymatic activity may have lasted millions, if not billions, of years, until by chance it created the RNA world, with its sophisticated and very flexible way of encoding genetic information.

With these speculations on the origins of life, we have reached the end of the story that started with Carleton Gajdusek and Vincent Zigas trekking the Highlands of New Guinea and wondering if cannibalism could be the cause of the spread of kuru among the Fore people.

To Recap

Mendel's laws predict the proportion of offspring that will inherit a given trait in various kinds of crosses between parents. They were discovered with plants but apply to all organisms with sexual reproduction.

Yeasts are simple eukaryotic organisms—that is, organisms with a cell nucleus. They are a favorite organism for geneticists. Surprisingly, some yeast traits do not follow Mendel's laws. This mystery was solved by Reed Wickner, who showed that such traits are caused by prion proteins.

Genes coding for prion proteins are found in all domains of life, even in viruses, as well as in archaea, bacteria, and eukaryotes. This suggests that they appeared early in the history of life on the planet.

Prion proteins have been retained during evolution because they are useful. They help organisms to deal with changes in their environment. Prion proteins causing disease are the exception. They are intrinsically disordered proteins prone to misfolding, which when misfolded become toxic for the cell. Their folding into prions is an accident with dreadful consequences.

Prion proteins may have been at the origins of life on the planet. This hypothesis is supported by some experimental evidence. The more commonly accepted hypothesis of an RNA world before the protein world is being challenged by the prion hypothesis

Epilogue

This book has taken us on a long journey through medicine and biology, from kuru and cannibalism in the Highlands of New Guinea to hypotheses on the origin of life on earth. Along the way, we looked at the structure of proteins and the way they fold. We considered a likely mechanism for self-templating, the process that can turn some proteins into prions. We described the role of such proteins in dreadful neurological diseases, including Alzheimer's and Parkinson's. We discussed the surprising discovery that some prion proteins play roles in essential functions such as long-term memory. We considered that self-templating proteins have been associated with life for hundreds of millions of years because of their usefulness, that they may even be at the origin of life on the planet.

This book was written for nonbiologists with an interest in science in general and neuroscience and medicine in particular. It belongs to the genre of "science communication," an activity that, I am convinced, is extremely important for society. One of its aims was to complement, and possibly amend, the type of science information that we find in the media. News stories tend to focus on remarkable discoveries, often announced as breakthroughs, without much discussion about how the results were obtained or, especially in medicine and therapeutics, how statistically significant they are. This has been the cause of much misunderstanding, sometimes with harmful consequences, such as vaccine refusal. Therefore, one of my goals was to give a more realistic picture, with some personal examples, of how things happen in the laboratory. How scientists study basic biological questions, such as mechanisms of memory, of diseases, and the possibilities of future treatments.

How do scientists go about solving the problems that they choose to study? I hope that I have convinced the reader that one of the most important steps is to clearly and carefully formulate the question that one wants to address. Next, to choose the best, which usually means the simplest, system to solve the question with reliable, reproducible, quantitative tools.

In medicine, choices may be limited. The medical investigator faces all the complexity of the disease, including variability from patient to patient, often the lack of an animal model, and so on. However, success here also depends on a clear formulation of the problem and on using appropriate quantitative tools. A first step is to isolate an important aspect of the disease that will be amenable to investigation. The question for the scientist is: Given what we already know, what is the most important aspect of the disease that now needs to be answered, and what tools are required to find the answer? Again, the key is to bring as much clarity as possible to setting one's goals. Once the question is clearly formulated, the scientist makes a hypothesis—a theoretical, plausible answer to the question—and then tests the hypothesis by performing experiments. A well-designed experiment should give a *yes* or *no* answer. If the answer is *no,* one needs to make a new hypothesis. If it is *yes,* one moves to the next level of complexity. This is obviously an ideal schema—the one that you present in your grant application and hope to be able to follow.

With luck, it will work as planned, but unexpected things often happen. Commonly, even the best designed experiment, instead of a *yes* or *no* answer, gives an entirely unexpected result that raises new questions instead of answering the original one. How disappointing! However, this is the juncture at which a good scientist reveals herself, using imagination. The research may take an entirely new direction—one risky but potentially fruitful, because it enters uncharted territory. Important discoveries have been made in this way.

Not all scientists like to follow a rational, Cartesian approach; there are different styles among researchers. One cannot ignore the role played by the personality of the scientist—and some personalities can be quite strong! Some proceed more haphazardly than I have described, testing one idea in one direction and at the same time another one in another direction, using intuition more than rational planning. If you are wondering which approach is best, the answer is *neither.* One should nurture diversity in research as in everything else. Variety of style is good. What is essential is to be always driven by curiosity. Research, even the most applied sort, remains a kind of a game, a game played against the unknown.

Pure curiosity is the essential ingredient of research; it is the fundamental reason for doing science. It is also very important for the physician. When seeing a patient for the first time,

"What is going on here?" is an instant reaction. This is what leads to an accurate, useful diagnosis. For the scientist too, "What is going on here?" is an everyday question, the one that drives him to design and interpret successful experiments. But on a more general level, curiosity in the face of a new, unexpected, bizarre observation, and the instinctive, possibly irrational need to find an answer, even when confronted with considerable difficulty, has led to many paradigm shifts in science.

In the case of prions, it was curiosity that sent Gajdusek and Zigas trekking the Highlands of New Guinea to investigate an exotic disease transmitted by cannibalism. It was curiosity that led Stanley Prusiner to embark, despite considerable skepticism and even scorn from many colleagues, on the identification of the "scrapie agent." Who would have predicted that kuru and scrapie would be responsible for the discovery of fundamental aspects of protein biology and of long-term memory? That this would open new vistas on the appearance of life on the planet billions of years ago? Nurturing curiosity, diversity of interests, far-fetched hypotheses—what science administrators like to call "high risk projects"—is of the utmost importance for medicine, for the patient, and of course for science in general.

For me, it also has been a long journey: from studying medicine at the time when Gibbs and Gajdusek announced that they had transmitted kuru and Creutzfeldt-Jakob disease to chimpanzees, to injecting mice with alpha-synuclein fibrils to test the prion hypothesis in Parkinson's disease, to teaching students about self-templating proteins. Obviously, medicine, and biology in general, has been through momentous changes during this period, changes with many beneficial consequences for patients. Looking back, what is most striking about these changes is that the pace at which they occurred increased considerably over the years. Just in the area of neurodegenerative diseases, the differentiation of induced pluripotent stem cells into neurons, the development of optogenetics, of PET scans, the ability to predict protein folding, powerful bioinformatics to dissect complex interacting mechanisms—all were developed simultaneously, or in rapid succession, over the last ten years. These are only technical advances, but they have opened new and powerful ways of studying disease mechanisms. As I have emphasized on several occasions, understanding diseases at the molecular level is the key to identifying drug targets and designing novel treatments. Clearly, the gap between finding a drug target and developing an efficacious, nontoxic drug is still wide and can be difficult to cross, as illustrated by the attempts to use antibodies to remove aggregated Abeta and tau from the brains of patients with Alzheimer's disease. But, while finding a cure for Alzheimer's and other neurodegenerative diseases is far from complete, I am convinced that the increasing pace at which biology progresses and the growing number of dedicated, bright scientists with diverse backgrounds interested in neurodegenerative diseases are together bringing the realization of efficacious treatments clearly into view.

ACKNOWLEDGMENTS

I was convinced by my wife, Beverley Bie Brahic, to turn my Stanford lectures into a book for nonscientists. She deserves credit and many thanks for the suggestion and her encouragement, which transformed a dreaded retirement into an exciting adventure. Several colleagues and friends read various versions of the text and offered valuable comments, suggestions, and corrections. For their essential contributions I wish to thank Robert Craft, Daniel Dunia, Dan Jarosz, Ronald Melki, René Monié, Neal Nathanson (in memoriam), and Jean-Pierre Roussarie. I thank Karla Kirkegaard for bringing me to Stanford after my mandatory retirement from Institut Pasteur, Aaron Gitler for many discussions on prions and for his hospitality in his laboratory, and Eric Freundt and Gregor Bieri for performing some of the experiments with alpha-synuclein described in the book. My heartfelt thanks to all the students who attended Stanford BIOS277 over the years and brought their enthusiasm for science and endless curiosity to the class. I wish to thank my editors: Ingrid Gnerlich for supporting my book project and for encouraging me to see it through, Whitney Rauenhorst for graciously answering all my questions and for solving many practical issues, and Maia Vaswani for superb copy editing.

GLOSSARY

Abeta. A fragment (peptide) of the neuron amyloid precursor protein. It aggregates and forms prion fibrils in the amyloid plaques of Alzheimer's disease.

Action potential. The weak electrical signal that travels from the dendrites to the cell body and then down the axon of a neuron that has been stimulated.

Aggregated proteins. Proteins that become insoluble and clump together. Clumps located inside cells make *inclusions*. Clumps outside cells form *plaques*.

AlphaFold. A computer program that predicts the folding of a protein from the sequence of its amino acids.

Alpha helices. A common pattern in the folding of proteins (see figure 2.1).

Alpha-synuclein. A neuron protein present at synapses. It folds into a prion form in Parkinson's disease.

Amino acids. Molecules that assemble in a linear chain to form a protein. There are twenty different amino acids.

Amyloid plaques. Aggregated proteins found in the brains of Alzheimer's disease patients. They are located outside cells and include fibrils of the Abeta peptide.

Astrocyte. A non-neuron brain cell that performs many different functions.

Axon. The single, long, slender extension of a neuron that carries action potentials away from the cell body. Some axons have multiple branching that connects them to many different secondary neurons.

Axonal transport. The mechanism by which molecules and organelles are transported long distances inside axons, either away from the cell body of the neuron or toward it.

Beta strands. A common pattern in the folding of proteins (see figure 2.1).

Channelrhodopsin. An ion channel that can be opened or closed depending on the kind of light it receives.

Chaperones. A class of proteins involved in protein quality control inside cells.

CPEB. A synapse protein that is induced into a prion form by repeated stimulation of the synapse. It plays a role in synapse remodeling and the acquisition of long-term memory. In the fruit fly the protein has been named Orb2.

Creutzfeldt-Jakob disease. A rare human disease with worldwide distribution caused by the prion form of the PrP protein.

Cryo-electron microscopy. A technique used to determine the structure of proteins. A protein solution is very quickly frozen and observed with a special electron microscope.

Dendrites. The many slender extensions of the cell body of a neuron. Dendrites receive nerve impulses from upstream neurons and transmit them to the cell body.

Dopamine. A neurotransmitter.

Electroencephalography (EEG). A technique that records the electrical activity of the brain from multiple electrodes positioned on the scalp.

Engram. The name given by Richard Semon to the network of neurons that is created during the acquisition of a specific memory.

Enzymes. Protein molecules that catalyze chemical reactions. They speed up chemical reactions that would be too slow (or may not happen at all) at body temperature.

Epigenetics. The study of heritable cell functions that do not depend on DNA.

Eukaryotes. Species whose cells have a nucleus limited by a membrane.

Functional magnetic resonance imaging (fMRI). A technique that maps the areas of the brain that are active when the subject performs a given task.

Genome-wide association study (GWAS). A genetic method to identify genes that confer a risk for a particular disease.

Glial cells. Non-neuron cells that perform many different functions in the brain. There are three types of glial cells: *astrocytes, oligodendrocytes,* and *microglial cells.*

Hippocampus. A structure deep in the brain that plays a central role in the acquisition and retrieval of memory.

Induced pluripotent stem cells (iPSCs). Stem cells that have been obtained by genetic manipulation of differentiated cells, such as skin cells.

Intrinsically disordered proteins. Proteins that do not spontaneously fold into a stable structure. They need to bind to a partner to fold properly.

Ion. An atom with an electrical charge. Atoms consist of a positively charged nucleus and a cloud of negatively charged electrons. The two cancel each other out so that the atom does not have a net charge. However, some atoms, like sodium, can lose an electron and become a positively charged ion. Others, like chlorine, can gain an extra electron and become a negatively charged ion.

Ion channel. A assembly of several proteins traversing a cell membrane. At its center is a pore that allows certain ions to cross the membrane.

Kuru. A human disease transmitted by cannibalism in the Highlands of New Guinea. It was caused by the prion form of the PrP protein. Kuru is now extinct.

Lewy bodies, Lewy neurites. The characteristic lesions observed in the brains of Parkinson's disease patients. They are made of aggregated proteins, including aggregated alpha-synuclein.

Lymphocyte. A category of blood cell, also present in the lymph nodes, spleen, and other organs, which plays various roles in the immune responses against pathogens. There are several kinds of lymphocytes, each with a specific function.

Mendel's laws. The rules of genetic inheritance that predict the distribution of characters (traits) of the parents that will be seen in the progeny of species with sexual reproduction.

Microglial cell. A type of non-neuron brain cell that detects foreign substances, such as microbes, and initiates an immune response against them.

Myelin. Layers of membrane wrapped around an axon. It isolates action potentials from the rest of the brain. It also provides nutrients to the axon.

Neurofibrillary tangles. Aggregated proteins found in axons in the brains of Alzheimer's disease patients. They include fibrils of the tau protein.

Neurotransmitter. A chemical substance secreted at a synapse by the upstream neuron. It binds to receptors on the downstream neuron, triggering a new action potential.

Nucleotides. The building blocks of nucleic acids. DNA and RNA are chains of nucleotides.

Oligodendrocyte. A non-neuron cell that forms the myelin sheath that surrounds many axons in the brain.

Optogenetics. A technique that allows neuroscientists to control the activity of neurons in the brain using light.

Orb2. *See* CPEB.

Organelle. A structure inside cells that performs a particular function.

Peptide. A short piece of a protein, usually less than a hundred amino acids long.

Plasma membrane. The membrane that separates a cell from its environment. Also called the cell membrane.

Positron emission tomography (PET). A technique of in vivo imaging to locate a particular protein in the body.

Prion. The original definition given by Stanley Prusiner in his 1982 article was: "Prions are small proteinaceous infectious particles which are resistant to inactivation by most procedures that modify nucleic acids" (*Science* 216, no. 4542 [1982]: 136–44). "Prion" sensu stricto refers to the misfolded, self-templating form of the PrP protein, responsible for spongiform encephalopathies in humans and animals.

Prion seed. A small fragment of a prion protein fibril that becomes elongated into a long new fibril by recruiting and folding more and more copies of the prion protein.

Promoter. A segment of DNA that controls the activity of a gene. Depending on the proteins that bind to the promoter, the gene can be turned on or off.

Protein fibril. A large assembly of prion proteins stacked on top of one another. They can be structures large enough to be visible under the electron microscope.

Protein folding. The series of reactions by which the linear chain of amino acids that makes a protein folds into a very specific shape required for the function of the protein.

PrP. A brain protein whose function is not clearly established. It can "misfold" into the prion responsible for scrapie and the other spongiform encephalopathies.

Ribosomes. Particles present inside cells made of protein and RNA. They are the micromachines that synthesize proteins.

RoseTTAfold. A computer program that predicts the folding of a protein from the sequence of its amino acids.

Scrapie. A sheep and goat brain disease caused by the prion form of the PrP protein.

Self-templating. The process by which prion proteins impose their prion folding onto other copies of the same protein.

Spongiform encephalopathies. *See* transmissible spongiform encephalopathies.

Stem cells. Cells that can differentiate into many cell types, including neurons. Stem cells are abundant in early embryos. Some stem cells are still present in the adult.

Substantia nigra. A small part of the brain, deep in the organ, whose neurons produce the neurotransmitter dopamine.

Synapse. A point of contact between two neurons. It is at synapses that the electrical action potential from upstream neurons is transferred to downstream neurons.

Synaptic remodeling. The physical enlargement that reinforces a synapse during the acquisition of long-term memory.

Tau. A neuron protein that aggregates and forms prion fibrils in the neurofibrillary tangles of Alzheimer's disease.

Transgenic animal. An animal, such as a laboratory mouse, in which one or several genes have been modified by methods of genetic engineering. Most often, an animal in which a foreign gene has been introduced into one of its chromosomes.

Transmissible spongiform encephalopathies. A group of fatal diseases caused by the PrP prion. The disease causes myriad microscopic holes in the brain, which under the microscope give it the appearance of a sponge, hence *spongiform*. Also known by the short form *spongiform encephalopathies*.

Viral vector. A genetically engineered virus used to deliver a gene to the cell that it infects. The virus has been mutated and cannot reproduce and cause disease.

X-ray diffraction. A technique used to determine the structure of proteins. Pure proteins need to be crystalized to be analyzed by X-ray diffraction.

INDEX